The Future of European Union Environmental Politics and Policy

The Future of European Union Environmental Politics and Policy investigates the trajectory of European Union (EU) environmental policy and reflects on how this hugely vital policy area of the EU has evolved over the decades.

Gathering together a selection of the leading scholars working on European environmental policy, the volume assesses the extent to which change has occurred in important dimensions of EU environmental policy research. These dimensions include the EU's values and approaches; the provision of leadership; the possibilities of Brexit and the dismantling of policies; policy instruments and climate change; policy implementation and enforcement; and policy evaluation. The contributors situate their research in the context of current developments and conditions, including the global economic challenges and the rise of political challenges to both European governance and integration. Each chapter reviews the EU environmental policy over the long term and assesses the implications of current developments for the future health of European environmental policy, European integration and the environment itself.

The Future of European Union Environmental Politics and Policy will be of great interest to scholars of environmental politics, environmental governance and EU policy. The chapters were originally published as a special issue of *Environmental Politics*.

Anthony R. Zito is Professor of European Public Policy at Newcastle University, UK. He is currently co-Director of the Jean Monnet Centre for Excellence at Newcastle University and co-Editor of the leading international environmental journal, *Environmental Politics*. His research focuses on governance, learning theory, networks and EU decision-making.

Charlotte Burns is Professor of Politics at the University of Sheffield, UK. She is co-Chair of the Brexit and Environment Network. Her research focuses upon EU environmental policy-making and decision-making and the impact of Brexit on EU/UK environmental policy.

Andrea Lenschow is Professor of European Integration and Politics and Director of the Jean Monnet Centre of Excellence in European Studies at Osnabrück University, Germany. Her research focuses on local, EU and global environmental policy and governance, policy integration, implementation and institutional change.

The Future of European Union Environmental Politics and Policy

Edited by
Anthony R. Zito, Charlotte Burns
and Andrea Lenschow

Routledge
Taylor & Francis Group
LONDON AND NEW YORK

First published in paperback 2024

First published 2020
by Routledge
4 Park Square, Milton Park, Abingdon, Oxon OX14 4RN

and by Routledge
605 Third Avenue, New York, NY 10158

Routledge is an imprint of the Taylor & Francis Group, an informa business

© 2020, 2024 Taylor & Francis

Chapter 5 © 2017 Charlotte Burns, Viviane Gravey, Andrew Jordan and Anthony Zito. Originally published as Open Access

Chapter 9 © 2017 Jonas J. Schoenefeld and Andrew J. Jordan. Originally published as Open Access

With the exception of Chapters 5 and 9, no part of this book may be reprinted or reproduced or utilised in any form or by any electronic, mechanical, or other means, now known or hereafter invented, including photocopying and recording, or in any information storage or retrieval system, without permission in writing from the publishers. For details on the rights for Chapters 5 and 9, please see the chapters' Open Access footnotes.

Trademark notice: Product or corporate names may be trademarks or registered trademarks, and are used only for identification and explanation without intent to infringe.

Publisher's Note
The publisher has gone to great lengths to ensure the quality of this reprint but points out that some imperfections in the original copies may be apparent.

Disclaimer
Every effort has been made to contact copyright holders for their permission to reprint material in this book. The publishers would be grateful to hear from any copyright holder who is not here acknowledged and will undertake to rectify any errors or omissions in future editions of this book.

British Library Cataloguing in Publication Data
A catalogue record for this book is available from the British Library

Typeset in Minion
by Newgen Publishing UK

ISBN: 978-0-367-46765-4 (hbk)
ISBN: 978-1-03-283941-7 (pbk)
ISBN: 978-1-00-303117-8 (ebk)

DOI: 10.4324/9781003031178

Contents

Citation Information		vii
Notes on Contributors		ix
Foreword David Judge		xi

1. Introduction: Is the trajectory of European Union environmental policy less certain?   1
   Anthony R. Zito, Charlotte Burns and Andrea Lenschow

2. Changing the story? The discourse of ecological modernisation in the European Union   22
   Amanda Machin

3. The importance of compatible beliefs for effective climate policy integration   42
   Katharina Rietig

4. The European Council, the Council and the Member States: changing environmental leadership dynamics in the European Union   62
   Rüdiger K.W. Wurzel, Duncan Liefferink and Maurizio Di Lullo

5. De-Europeanising or disengaging? EU environmental policy and Brexit   85
   Charlotte Burns, Viviane Gravey, Andrew Jordan and Anthony Zito

6. Voluntary instruments for ambitious objectives? The experience of the EU Covenant of Mayors   107
   Ekaterina Domorenok

7 Compliance with EU environmental law. The iceberg is melting 129
   Tanja A. Börzel and Aron Buzogány

8 Left to interest groups? On the prospects for enforcing
   environmental law in the European Union 156
   Andreas Hofmann

9 Environmental policy evaluation in the EU: between learning,
   accountability, and political opportunities? 179
   Jonas J. Schoenefeld and Andrew J. Jordan

Index 199

Citation Information

The chapters in this book were originally published in *Environmental Politics*, volume 28, issue 2 (March 2019). When citing this material, please use the original page numbering for each article, as follows:

Chapter 1
Introduction: Is the trajectory of European Union environmental policy less certain?
Anthony R. Zito, Charlotte Burns and Andrea Lenschow
Environmental Politics, volume 28, issue 2 (March 2019), pp. 187–207

Chapter 2
Changing the story? The discourse of ecological modernisation in the European Union
Amanda Machin
Environmental Politics, volume 28, issue 2 (March 2019), pp. 208–227

Chapter 3
The importance of compatible beliefs for effective climate policy integration
Katharina Rietig
Environmental Politics, volume 28, issue 2 (March 2019), pp. 228–247

Chapter 4
The European Council, the Council and the Member States: changing environmental leadership dynamics in the European Union
Rüdiger K.W. Wurzel, Duncan Liefferink and Maurizio Di Lullo
Environmental Politics, volume 28, issue 2 (March 2019), pp. 248–270

Chapter 5
De-Europeanising or disengaging? EU environmental policy and Brexit
Charlotte Burns, Viviane Gravey, Andrew Jordan and Anthony Zito
Environmental Politics, volume 28, issue 2 (March 2019), pp. 271–292

Chapter 6
Voluntary instruments for ambitious objectives? The experience of the EU Covenant of Mayors
Ekaterina Domorenok
Environmental Politics, volume 28, issue 2 (March 2019), pp. 293–314

Chapter 7
Compliance with EU environmental law. The iceberg is melting
Tanja A. Börzel and Aron Buzogány
Environmental Politics, volume 28, issue 2 (March 2019), pp. 315–341

Chapter 8
Left to interest groups? On the prospects for enforcing environmental law in the European Union
Andreas Hofmann
Environmental Politics, volume 28, issue 2 (March 2019), pp. 342–364

Chapter 9
Environmental policy evaluation in the EU: between learning, accountability, and political opportunities?
Jonas J. Schoenefeld and Andrew J. Jordan
Environmental Politics, volume 28, issue 2 (March 2019), pp. 365–384

For any permission-related enquiries please visit:
www.tandfonline.com/page/help/permissions

Notes on Contributors

Tanja A. Börzel is Professor of Political Science and holds the Chair for European Integration at the Otto-Suhr-Institute for Political Science, Freie Universität Berlin, Germany. Her research focus and teaching experience lie in the field of institutional theory and governance, comparative regionalism and diffusion.

Charlotte Burns is Professor of Politics at the University of Sheffield, UK. She is co-Chair of the Brexit and Environment Network. Her research focuses upon EU environmental policy-making and decision-making and the impact of Brexit on EU/UK environmental policy.

Aron Buzogány is a professor in the Department of Social Sciences in the Institute of Forest, Environmental and Natural Resource Policy (InFER) at BOKU University of Natural Resources and Life Sciences, Vienna, Austria.

Maurizio Di Lullo is a senior advisor at the Council of the European Union, Brussels, Belgium. He is a lawyer and linguist by training and has a decade of experience in international climate negotiations.

Ekaterina Domorenok is Associate Professor of Political Science at the University of Padua, Italy. Her research interests mainly concern policy design, implementation and learning in multi-level settings, with particular regard to EU policies for climate, sustainability, environment and cohesion.

Viviane Gravey is a lecturer in the Department of Politics and International Studies at Queen's University Belfast, UK.

Andreas Hofmann is a researcher at Freie Universität Berlin, Germany, where he is a member of the Centre for European Integration at the Otto-Suhr-Institute for Political Science. His primary research interest is the role of courts in political processes and judicial procedures as a mode of solving political conflicts.

Andrew J. Jordan is Professor of Environmental Sciences in the Centre for Social and Economic Research on the Global Environment (CSERGE). He is fascinated by the politics that emerge when attempts are made to govern environmental problems using the tools and methods of public policy.

Andrea Lenschow is Professor of European Integration and Politics and Director of the Jean Monnet Centre of Excellence in European Studies at Osnabrück University, Germany. Her research focuses on local, EU and global environmental policy and governance, policy integration, implementation and institutional change.

Duncan Liefferink is an assistant professor in the Environmental Governance and Politics group, Institute for Management Research, at Radboud University Nijmegen, the Netherlands.

Amanda Machin is Professor for International Political Studies at the University of Witten/Herdecke, Germany. Her research focuses particularly on environmental politics.

Katharina Rietig is Lecturer in International Politics at Newcastle University, UK. Her research sits at the intersection of international relations/global governance and European public policy.

Jonas J. Schoenefeld is a research associate at the University of East Anglia, UK. He is pursuing a PhD on climate policy evaluation with Professor Andrew J. Jordan at the Tyndall Centre for Climate Change Research in the School of Environmental Sciences.

Rüdiger K.W. Wurzel is Professor of Comparative European Politics and Jean Monnet Chair in European Union Studies at the University of Hull, UK, where he is Director of the Centre for European Union Studies (CEUS).

Anthony R. Zito is Professor of European Public Policy at Newcastle University, UK. He is currently co-Director of the Jean Monnet Centre for Excellence at Newcastle University and co-Editor of the leading international environmental journal, *Environmental Politics*. His research focuses on governance, learning theory, networks and EU decision-making.

Foreword

Life is full of surprises. One of which was to be asked to write a foreword to this volume. Nearly 30 years ago I had been asked by *Environmental Politics* to edit the first of its special issues in the very first year of its publication in 1992. That special issue subsequently appeared in book-form as *A Green Dimension for the European Community: Political Issues and Processes*. I was surprised to find, therefore, despite the passage of time, that *A Green Dimension* had not been confined irretrievably to the 'remainder bin' of publishing history, and, instead, had served as a prompt for the editors of the present volume to produce an incisive, systematic review of present-day European Union (EU) environmental policy and governance.

A Green Dimension was a product of its time. It was commissioned as the target date for the 'completion' of the Single Market drew close – in accordance with the provisions of the Single European Act (SEA). Significantly, the SEA's economic provisions sat alongside institutional innovations, notably the cooperation procedure and the extension of qualified majority voting, and explicit treaty-based recognition of environmental policy competences. The timing of 1992 thus provided an opportunity to bring together a number of scholars – working respectively in the fields of environmental politics, political economy, policy analysis, electoral and participatory politics – to consider the interconnectedness of emergent environmental policies and political and policy processes within the, then, European Community (EC). To guide this investigation, the contributors were offered 'directional pointers' to examine how environmental issues were framed, articulated politically, and translated into EC policies in the interplay of interinstitutional bargaining and multi-level politics. In this respect, the resultant volume was multi-dimensional in its analysis of environmental politics and policy in the EC, rather than focusing upon a single dimension as suggested in its title.

Equally, *The Future of European Environmental Politics and Policy* is a product of its time. But the times of the 2020s are radically different to those of the 1990s. Whereas *A Green Dimension* was focused upon emergent perspectives, policies, and political interactions and interventions, the present volume is focused upon the nature and extent of political and policy *change*. A new set of 'directional pointers' have been provided by the editors to cohere analyses of the changing dynamics of environmental discourses and ideational framing, of the configurations and reconfigurations of inter-institutional relationships and multi-level governance arenas, and of evaluations of environmental policies and programmes. In following these 'pointers', the eminent contributors reveal not only the complexities of theorisations, conceptualisations, discourses and empirical analyses of EU environmental policies and policy-making but provide sophisticated expositions with which to make sense of those complexities. For this alone this volume provides a benchmark for the contemporary study of environmental discourse and governance within the EU. But this volume is much more than a reflection of intellectual acuity. In a time of multiple interlinked 'crises' – at global, EU, and state level – it offers astute, grounded assessments of the resiliences, capacities and inhibitors inhered within EU environmental policy-making. In uncertain times this volume offers the reader a certainty of judgement; both in its chronicling of changes in environmental policy and governance and in its assiduous analysis of the significance of those changes within and beyond the EU.

David Judge
Professor Emeritus of Politics at the University of Strathclyde

Is the trajectory of European Union environmental policy less certain?

Anthony R. Zito ⓘ, Charlotte Burns ⓘ and Andrea Lenschow ⓘ

ABSTRACT
The core themes and research questions of this volume, centred on the nature of environmental policy change in the European Union (EU), are laid out. An original heuristic framework to capture different dimensions, mechanisms and processes of policy change is presented. In order to contextualise the current situation, where EU policy scope has reached maturity and faces an uncertain future trajectory, EU environmental politics is divided into particular eras, looking closely at the nature of change in each period. This volume interrogates the extent to which change has occurred, the conditions or context within which it did/did not take place and the implications arising from stasis or change. The contributions to the volume are introduced and placed into the context of the broader trajectory of EU environmental policy.

Introduction

Since the 1992 *Environmental Politics* Special Issue on European Community Environmental Politics (Judge 1992), it is striking that no journal has produced a systematic, cutting edge survey of European Union (EU) environmental politics and policy. This volume rectifies that omission.

The 1992 Special Issue was published at a time in which the European Single Market was creating renewed optimism about European integration (Weale and Williams 1992), and the Brundtland Commission was placing environmental action at the political cutting edge (Brundtland *et al.* 1987). The contributors to that Special Issue explicitly linked this momentum towards European integration with environmental policy (Hildebrand 1992; van der Stratten 1992; Weale and Williams 1992). However, while overall EU integration and environmental protection have moved in parallel and have been mutually reinforcing, they are nevertheless two distinct processes, varying in speed, intensity and accomplishment. Consequently, we cannot presume as to how European integration and environmental politics and policy will interact. Equally, we acknowledge the importance of environmental policy for the integration project (Lenschow and Sprungk 2010): EU elites used environmental policy, the perception of its success and the policy's popularity to legitimise EU integration.

This discussion raises several analytical questions. What is the nature of change in EU environmental policy, and what factors have driven these changes? If EU environmental policy is declining, stagnating or increasing, what does this tell us about wider EU integration? Correspondingly, if European integration is struggling, what are the implications for the EU's environment sector?

Despite our focus on the environment, it is important to acknowledge the uncertainties at work in EU integration as it shapes European environmental politics and policy. EU studies have focused recently on 'post-functionalist' perspectives (there is no longer an expectation that EU integration will ratchet in one direction, e.g. Hooghe and Marks 2009) and crisis (e.g. Laffan 2016). The economic, political and social difficulties (including the Euro crisis) confronting the EU have contributed to differentiated integration between member states (meaning here that certain EU rules may only govern a subset of member states or that individual state implementation varies: Holzinger and Schimmelfennig 2012), and arguably the first case of EU political disintegration (Webber 2014), namely 'Brexit'.

Simultaneously, global dynamics are challenging EU environmental politics and policy. Other global priorities have pushed the environment down the policy agenda: economic recession, migration, managing conflict and concerns over energy security. The nature of contemporary environmental problems has also become more challenging. While policy governance has dealt with the easiest environmental policy problems, the more politically intractable problems of climate change, overconsumption and diffuse pollution remain. These longer term challenges clash with the reality of electoral politics, where political attention focuses on a much shorter politics and policy time horizon. All these circumstances challenge the ability of the governing elite and the Western democratic state to govern, and to adopt and implement regulatory policies protecting the environment. As these environmental challenges are common to all OECD states, examining the EU has much wider resonance.

Analytical focus

The 1992 Special Issue of *Environmental Politics* focused on political institutions (e.g. the Single Market), political actors (e.g. European Parliament

Environment Committee, environmental groups and the green political movement), issues (e.g. reconciling economic interests and classical economics with environmental values and concerns) and processes (e.g. implementation) that were pushing European environmental integration forward. We focus on a similar range of factors but analytically emphasise questions about the nature of political and policy *change*. We examine how different elements of the EU environmental arena are responding to or driving change. We develop an original heuristic framework for assessing the range of environmental policy change.

This volume brings together leading academics working on the central aspects of EU environmental policy research. Each author addresses a set of core questions:

- Has there been change?
- How significant or extensive has that change been?
- What has been the key driver of or obstacle to change?
- Does the key driver or obstacle to change exist within the EU environmental political realm or does it operate externally to the sector and perhaps even to the EU?
- What does change mean for our understanding of EU environmental politics/policy?

After describing the heuristic framework in the next section, we deploy this framework to briefly trace and reinterpret the evolution of EU environmental policy. We discern four distinctive eras of EU environmental politics and policy: the first (1967–1984) focuses on the initial experimentation and the rise of social movement interest; the second (1985–1999) covers the growth of environmental politics operating in a context of accommodating the market and the ideology of neoliberalism; the third (1999–2008) sees the growing recognition of the reality of intractable problems and the questioning of the regulatory state; the fourth (post-2008) period encompasses the current challenges facing the EU. In the discussion of the fourth era, we introduce each contribution to the volume, using the findings to assess the EU's future trajectory.

Theoretical approaches to change

We offer a heuristic framework for understanding change (summarised in Table 1), which, generated from a range of literature, highlights an array of possibilities without the expectation that our authors, who will also pursue their own theoretical concerns, employ all of them. In order to capture the range of possible change dynamics, our framework speaks to two distinct dimensions that do not necessarily correspond to each other, the key

Table 1. Analytical framework for understanding policy change.

Dimensions of change	Process dynamics of change
How do actors view and assess the world? Learning and discursive changes.	*No change* Status quo is preserved.
Which actors occupy the political decision-making arena? Changes in the dominant elite, government, network and/or coalitions in the arena.	*Incremental change* Status quo tends to remain in place over time but accumulation of change can occasionally overthrow it.
How has the arena changed? Changes in institutions, operating culture, political opportunity structures.	*Punctuated equilibrium* Status quo largely stable except for particular moments when significant change is possible.
	Rapid and repeated episodes of disequilibria Multiple challenges to the old status quo and the creation of a new one.

dynamics that are involved with change in the system; and how the process of change develops, and whether it represents continuity or discontinuity compared to the *status quo*. To avoid proliferating new concepts and jargon, we synthesise our framework using already extant analytical constructs. Given the vast theoretical literature on political and policy change we acknowledge only some of the more prominent and useful works. Our framework shares some characteristics of Capano and Howlett's (2009) special issue on policy change, but differs in significant aspects (i.e. taking a more eclectic approach to the theoretical approaches assessing the change dynamics and concentrating on how the actors within the EU environmental sector have had to interpret and respond to external circumstances [Real-Dato 2009]).

Dimensions of change

We derive three essential questions from the literature on political and policy change: how do actors view the world while operating in the EU environmental policy area; which actors control the key political arena; and how have the parameters of this arena changed? A considerable number of the analytical frameworks combine elements of these dynamics (see below).

A range of explanations focus on how ideational and discursive dynamics shape politics. On the more empiricist and behaviouralist side are studies of learning and lesson drawing. Learning in these various manifestations involves individuals or groups making a judgment based on an experience, a new way of thinking or some other input inducing actors to articulate a different view of how things happen and how actors should respond (May 1992; Zito and Schout 2009). On the post-empiricist side is the focus on discourse and other elements of power. Hajer (2005, p. 299) defines the discursive approach as how the 'definition of a problem relates to the particular narrative in which it is discussed'. Influenced by thinkers such as Foucault, these approaches often focus on the manner in which long-term social structures arise out of human interactions, and the means by which power can act at a distance (Latour 1986, p. 264). Applying some EU examples, the former approaches capture how European thinking about how acid rain and its mitigation has shifted over time, whilst the latter would focus on the language and discourse surrounding sustainable development or on the narrative about the EU's central role in environmental policy (Lenschow and Sprungk 2010).

Some prominent approaches share elements of both tendencies, such as policy framing where actors confront situations where their understanding is uncertain and problematic, and develop insights to enable assessment and action (Schön and Rein 1994). Also notable are neo-Gramscian arguments about dominant ideologies governing what is possible, and limiting the potential for change so as to benefit particular elites (Levy and Newell 2002).

In contrast to ideational/learning/discursive approaches are those that assess who controls power within a political arena. Schattschneider (1975) argued that politics was the question of who was involved in the processes of conflict: those in control of the arena restrict the number of other political actors involved in making political choices. Those political actors inside the arena dissatisfied with the political *status quo* and those actors marginalised from decision-making have an interest in getting the 'scope of conflict' expanded by bringing in new voices or holding elections enabling a wider range of views to shape and disrupt the political arena. For example, Donald Trump's election has affected both United States (US) environmental policy and the global environmental arena in which the EU operates. While Schattschneider focused on group theory, the institutionalist argument about who controls the political system's 'veto points' raises related concerns (Tsebelis 1990). This approach asks which actors in the decision-making process have the ability to pass or block initiatives using powers given by the system. Changes in who controls veto points may affect policy outputs. Thus, granting the European Parliament co-legislative powers with the Council of Ministers made the Parliament a stronger veto player in EU environmental policy.

If we move from the approaches concerned with overarching systems, considerable literature focuses on party competition and changes in government (e.g. Brady 1978); the rise of the German Greens and its effect on German and other European electoral politics is one important example. Much public policy literature examines the role of policy networks. Such analysis often speaks to two dimensions, the degree to which a particular network controls a decision-making process, or more internal parameters such as the position of individual actors within the network (Rhodes 1990). These networks may coalesce informally/formally, as in the EU Network for the Implementation and Enforcement of Environmental Law. Other scholars have focused on the presence of political and policy entrepreneurs who advocate a particular policy or political position and invest skills, knowledge and other resources to bring about change (Kingdon 1984). In the EU context, scholars have credited the European Commission, specifically DG Environment officials, with championing particular policy problems and solutions.

Network and party analyses often (but not always) focus on national and subnational dynamics, but other approaches also study transnational systems such as the EU as a political arena. (Moravcsik's 1993) liberal intergovernmental approach to European integration studies how a dominant national set of political and economic actors shapes a member state's approach to EU negotiations. Depending on the power of the interested states, and the degree of interest each member state takes, certain states will drive this arena and the direction of integration. The discussion about which states are EU environmental leaders, such as Denmark, taps into this discussion (see Wurzel *et al.* 2019 – this volume).

The last explanation of change considered here concerns changes to the actual structure and rules of the game. To save space, we discuss only two dynamics, institutionalist and cultural. March and Olsen (1989) refocused political science on the structures of rules and norms shaping actor behaviour over time. Historical institutionalist models focus on the incentives institutions create rendering radical change difficult to adopt. At the start of the process, there are a number of potential outcomes, but seemingly minor events, as long as they occur at a propitious moment, can have lasting and, sometimes, self-reinforcing effects that normally inhibit policy reversals (Pierson 2000).

March and Olsen's perspective, particularly the logic of appropriateness (where institutional norms steer how actors view the world and behave), shares aspects of a cultural approach. Thompson *et al.* (1990, pp. 2–5) suggest that human cultures limit the range of possibilities for how norms drive behaviour. How individual members behave reflects the degree to which the wider societal group incorporates the individual (greater incorporation giving greater scope for the group to define the individual's choices) and the degree to which externally imposed prescriptions govern the individual. Expectations of both institutional and cultural approaches are that substantive change is less likely in the short term, but the implications of that change may become profound over time. An example is the path dependent impact of the Common Agricultural Policy in creating institutionalised decisions and cultural norms that have proved difficult for green values and priorities to modify (Lenschow and Zito 1998).

Process dynamics of change

Our second array of dynamics focuses on the process of change and the degree to which the *status quo* changes. To assess the degree to which the *status quo* changes we should analyse the directionality of that change and also whether changes accumulate towards a new and different equilibrium (Howlett and Cashore 2009, p. 41).

The first framework scenario (Table 1) is where no change happens; the *status quo* remains. This is not the same as saying that there is no politics. There may be a strong exertion of political power and ideas to maintain the *status quo* for a given set of political and policy choices. Incorporating ideational thinking, Edelman (1985) suggested that much of democratic politics, certainly in the US context, reflected an effort by the political elite to shape public perceptions of politics by often deploying empty symbols. A key dynamic was the deployment of symbolic language and appeals to emotion to distort or conceal political realities. The scenario we focus on is one where actors evoke symbols, rhetoric and discourses with no actual goal and prospect of change, often to obscure and conceal the lack of change. The EU's effort to create a 'green' agriculture in the 1980s and 1990s is one example where the rhetoric of a green reform has been stronger than the actual mitigation of negative environmental externalities of the Common Agricultural Policy (Lenschow and Zito 1998). A different scenario with the same result is the notion of cyclical patterns of change, where substantial change can occur but with an eventual return to the *status quo* over the longer term (Capano 2009).

The second process of change is incremental, suggesting that change happens gradually. This volume demonstrates that there is an analytical question about whether incremental change is likely to lead to a significant change in the *status quo* and when. Certainly, significant change is possible but it more likely happens over a substantial time-period.

An influential explanation for how European integration might proceed is neofunctionalism. Haas (1958) hypothesised that European integration followed gradual, ratcheting trajectories involving self-reinforcing dynamics. He adopted the term 'spillover' to describe the dynamic; successful policy achievements in one area of integration create a political recognition of the value of the achievement (and consequently of the organisation that produced this political good), and the incentive/political pressure for actors to enhance this extant integration by extending it into new areas. Linking this analysis to the first part of our framework, part of the causal explanation of the neofunctional approach is a learning dynamic, but certain institutional actors and economic interests also figure, using their influential positions to drive this process.

Over time, neofunctional change is likely to be substantive in transforming politics within the system, but may not necessarily do so at each spillover movement. The focus of neofunctionalism is on incremental steps, but certain of these steps may be significant in their own right (e.g. the EU Emissions Trading Scheme adoption within EU climate change policy). Significantly, this perspective allows for the possibility of integration reversing itself; the term 'spill-back' involves actors retreating from integration *status quo* and diminishing the scope of integration (Schmitter 1970).

As noted above, especially historical and sociological institutionalist perspectives focusing on path dependency and the logic of appropriateness (which is rooted in deeply anchored normative structures and traditions) tend to see change occurring gradually in a way that does not alter the *status quo*. Nevertheless, these institutionalist schools see scope within such a mechanism for substantial, cumulative change, although the expectation is that it occurs gradually over an extensive time-period. Streeck and Thelen's (2005, pp. 18–31) typology outlines different scenarios in which the gradual evolution of institutions may nevertheless generate significant, long-term political/policy change. They isolate five potential phenomena: displacement (one minor element in the system ends up dominant), layering (additional elements are added to an existing system over time and change its nature), drift (elements of the system gradually deteriorate over time due to deliberate neglect), conversion (transformation of an existing system to a new purpose involving active redirection) and neglect (the system's gradual disintegration).

The punctuated equilibrium concept implies that political systems generally maintain stasis over time, but that there are occasions/crises when changes occur producing 'large-scale departures from the past' (True *et al.* 1999). Political and policy choices and events follow on from past practice, usually incrementally (Baumgartner and Jones 2010). Nevertheless, circumstances occasionally occur where the current political and policy paradigm for governing and governance is challenged, potentially opening a wider range of available choices. The causal dynamic is likely to involve changes in who controls the decision-making process but also changes in paradigms and ideas. These moments of punctuated equilibrium trigger greater interest from political actors and the general public in what happens in specialised sectors such as the environment (expanding the scope of conflict). There is potential for sharp changes in political and policy direction, although those seeking to uphold the *status quo* will fiercely resist them (True *et al.* 1999). This change in the *status quo* may seem abrupt even if challenges have been building over time, e.g. German energy policy changed significantly after the 2011 Fukushima disaster, but debates and proposals in Germany concerning nuclear power had existed decades previously.

The most extreme change involves the possibility that a range of events occurs that destabilises the political process and policymaking. Focusing more on cultural dynamics, Eckstein (1988, p. 794) described the possibility of the persistence of novel situations that may lead to institutional and policy discontinuity. Contextual changes (e.g. global economic crises, national emergencies) could be so 'considerable or rapid or both' that political systems are unable to deploy their *status quo* maintenance or to be sufficiently flexible to have the *status quo* adjust to this change, despite strong societal desire to maintain the stability of their perspectives and values (Eckstein 1988, p. 796). Examples include cases of rapid industrialisation with the resulting political, economic and social consequences, or the situation facing interwar Germany where war, the peace settlement and economic traumas created shattering political consequences (Huntingdon 1971; Eckstein 1988). The buffeting change leads to situations where it takes time for a society to establish a new continuity, often at considerable social cost.

Findings: the eras and trajectories of EU environmental policy

The next four subsections provide a new analysis of EU environmental policy's evolution, its stages and various political and policy changes. We organise each time-period around significant types of policy changes, making use of Hall's 1993 framework. We focus on changes to: policy paradigms (major, substantive change or transformation in political values and their prioritisation contained in the policy approach), policy programmes (array of instruments and approaches responding to particular political issues and concerns), and selection of particular policy instruments/tools that achieve policy goals. Inherent in this approach is the distinction between evolutionary and revolutionary changes (Capano and Howlett 2009). The scholarly expectation is that changes at the micro (instrumental) level will be more evolutionary, but we do not rule out the potentially revolutionary role of instruments such as the EU Emissions Trading Scheme.

The time-periods are more indicative than fixed as many dynamics and trends of one era leach into succeeding eras, making hard and fast determinations analytically difficult. Our analysis differs from Hildebrand (1992), McCormick (2001) and Knill and Liefferink (2007) in that it compares eras systematically using concepts of change. Second, similar to Delreux and Happaerts (2016), we explicitly acknowledge the global linkages and processes that EU environmental politics and policy have shared.

Establishment and early growth: 1967–1984

Although Hildebrand (1992) and others reasonably assess EU environmental policy from the original 1957 treaty formation, we view the European Communities (EC) environmental programme as requiring a significant framing of the environment as a political/policy problem. The 1967 wreck of the oil tanker *Torrey Canyon*, impacting the UK and France, contributed to a European awareness of the pollution problem. Such events punctuated the political equilibrium, giving momentum to a worldwide social movement in advanced industrial states to frame the environment as a problem. Diverse governments, such as the conservative Nixon White House joined European governments such as the West German Brandt coalition in co-opting or embracing the global progressive politics including environmental protection, and creating legislation and institutions. The Commission and EC level processes shared this shift in world-view but had a primary political aim of protecting and enhancing the Common/Single Market. The spillover dynamic of finding linkages between extant integration and new policy areas asserted itself as the Single Market created pressures to harmonise potentially diverging national legislation and to address the threat of national environmental legislation becoming de facto barriers to trade and member state competitiveness (Weale 1999). Simultaneously, however, EC treaties gave EU environmental policy its foundations, with significant institutional implications. The path dependent linkage of environmental policy to Single Market dynamics (for a substantial part of its political impetus and the bulk of its legal justification through EC Treaty Articles 100 and 235) meant that the imperatives of the single market framing also featured in environmental policy (Rehbinder and Stewart 1985; Hildebrand 1992).

Thus, the early 1970s reveal two processes of change colliding in a way that boosted environmental policy: the integration spillover dynamic of the Single Market intersected, influenced, and was influenced by a punctuated equilibrium dynamic that reached its disequilibrium moment in 1972–1973. By the early 1970s, EC decision-makers in the Council and Commission were conscious not only of the social discourse on environmental concerns but the increasing potential spillover implication of member state legislation coming into force. This culminated in a strong shift in the EC political and policy programme without having explicit mention in the original EC treaties.

From 1973 to 1984, the EC environmental programme focused on developing legally binding acts (200 by 1985 according to Knill and Liefferink 2007). This reflected the lack of alternative tools (e.g. very little budget to try incentives such as subsidies and no power to create environmental taxes) and the reality that much of the environmental field was steering member state environmental protection around the protection of the Common Market. The main regulatory instrument was the directive, giving EC member states scope to tailor how they achieved the directive objectives to specific domestic legal and policy circumstances. As with 1970s member state legislation, EU policy had a focus on specific environmental media (i.e. the components of the natural environment) regulations, focusing on air and water pollution in particular (Wurzel 2002).

Reconciling market and environmental policy impulses: 1985–1999

Ratified in 1986, the Single European Act (SEA) had generated by 1985 a political excitement for the European integration project with profound implications for all EC areas, including environmental policy. There was a surge in the spillover dynamic where a confident Commission and member state governments shared an interest in enhancing member state economic performance. The SEA explicitly recognised the environmental policy aims with three new articles, reflecting an independent political valuation of environmental protection as an important EC goal (European Communities 1987). High profile 1980s international diplomatic efforts helped generate another significant legal recognition, with the EU gaining status in international environmental negotiations, separate from recognition given to member states (Vogler 1999; Delreux and Happaerts 2016).

Equally important was the idea of sustainable development, integrating economic growth with environmental concerns, gaining visibility through the Brundtland Commission and other efforts (Brundtland et al. 1987). The continued desire to fulfil both ambitions led to a sometimes explicit, often implicit effort to reconcile environmental protection with neo-liberal prioritisation of market values and solutions and concern about the consequences of regulatory approaches. Environmental damage became framed as a distortion within the market with environmental costs needing to be internalised in the market (Jachtenfuchs 1996). This framing gave greater scope to consider other forms of environmental policy instruments, but also safeguarded neo-liberal assumptions (Machin 2019 – this volume). Nonetheless, the EU repertoire of policy tools largely remained binding regulatory legislation. Efforts to create a carbon-energy tax failed (Zito 2000). Information campaigns played a comparatively minor role although there were successful programmes including the ecolabel Blue Flag (Blue Flag 2016) and creation of the European Environment Agency.

Challenging 'normal' environmental governance: 1999–2008

In 1999, 11 member states adopted the Euro as their common currency. Although EU environmental goals continued to be touted loudly, sometimes prominently, in such negotiations as Kyoto and its ratification efforts (promoting the image of the EU as a key global environmental leader, Oberthür and Dupont 2011), after 1999 the priority of safeguarding national economies and their competitiveness shifted the environment down the agendas of the EU and member states. The 2000 Lisbon Process enshrined sustainability in its goals, but the Process' evolution over a 10-year period is telling evidence of an increasing focus on economic achievements and environment's gradually decreasing prominence (Interview, Commission official, 10.1.17).

1999 is also the year that EU Commission President Santer and fellow Commissioners resigned, denting the Commission's prestige, creating a leadership vacuum and shift in policy focus (Cini 2008). Much of the scandal focused on whether the EU could manage programmes effectively in the wake of the Delors Commission's integration expansion. Thus, the EU's claim to legitimacy through the creation of effective public, including environmental policies, was open to challenge (Knill and Lenschow 2002). The Commission's reaction was not to row back on environmental policy, but its ambitions were reigned in and there was greater focus on improving processes (Interview, Commission official, 10.1.17). During this period, the EU also faced the great political and economic demands of integrating a large body of new, often poorer states; juggling closer economic convergence with expansion shifted the EU's focus and the attention of EU processes.

There is no evidence of punctuated equilibrium, discontinuity or spill-back changes in this process as there was no strong redirection, reversal or spill-back of environmental policy. The function of valuing and protecting the environment did not change as an EU priority. EU organisations, programmes and policy instruments remained in place; furthermore, the EU could highlight some global leadership efforts especially in climate change but also in areas such as biosafety (Delreux and Happaerts 2016). This outcome suggests a process of incremental change where other EU priorities displace environmental protection, which perhaps loses direction. The annual adoption rate of EU environmental legislation dropped between 2002 and 2007, recovered somewhat and then dropped again in the 2010s (Haigh 2011; Wurzel et al. 2013).

There was also a push to make regulation less onerous, and more flexible. The Commission proposed less intrusive legislation such as framework directives that gave greater scope for member state implementation, and non-legislative instruments based on the principle of shared responsibility (Jordan 1999). Since the mid-1990s, all Commission environmental proposals are required to incorporate a cost-effectiveness statement (Wurzel et al. 2013).

The uncertain future: 2008 and beyond

In August 2007, BNP Paribas made its hedge fund announcement (Dealbook 2007); the global financial crisis and resulting economic crisis soon followed. This was not the only challenge facing the EU: in 2008 Russian actions in the breakaway republics in Georgia, a country with which the EU was building closer ties, created a new security and foreign relations reality for the EU. In relatively quick succession, the European Sovereign Debt Crisis from 2009 onwards was followed by Syrian refugee migration starting in 2011, and the UK voted to leave the EU in 2016.

Of the myriad issues these events raise, we concentrate on three. First, the events challenge the direction and fundamental viability of the

European integration project across political, social and economic grounds. Second, it is a long-term empirical question as to whether this represents a period of disequilibria (Eckstein 1988) that will fundamentally change the EU's nature. Third, whilst still speculative it is also timely to assess how far such disequilibria affect the integration process and the EU environmental sector. A growing scholarship suggests that the EU may not be dismantling its environmental policy, but there is a greater rhetorical drive to 'fight red tape' and moderate environmental ambitions that may have long-term consequences (Burns and Tobin 2016; Gravey and Jordan 2016; Steinebach and Knill 2017). The contributions assembled here explore the nature and trajectory of EU environmental policy in this more uncertain context.

The contributions introduced from different starting points, the contributors engage with all three of these aspects of public policy, the framing role of ideas and discourse, institutional actors and arenas, and the evolution of policy implementation in the EU context.

Machin (2019) focuses on the evolution of ecological modernisation, one of the core ideological principles and discourses in the EU, exploring how ecological modernisation at the paradigmatic and instrument levels, as made manifest in EU strategic documents and the prototypical environmental policy instrument (the Emissions Trading Scheme), has evolved incrementally towards giving pre-eminence to the market's role and position. In doing so, this discourse constrains politics and excludes alternative ideas to improve policy and policy implementation.

Also focusing on more intangible qualities that drive change, Rietig (2019 – this volume) examines how the EU Commission has shaped its climate change strategy, focusing explicitly on the Commission's entrepreneurial effort, arguing that it has seized advantage of external opportunities and compatible beliefs across policy sectors to develop a substantial EU renewable energy policy programme. When the Commission satisfied external international commitments and exploited compatible policy values in the other related policy sectors and general societal receptivity, rapid spillover occurred, producing the Renewable Energy Directive. The absence of these external conditions and lack of agreement within the Commission leadership limited the scope for Commission entrepreneurship in the area of climate policy integration with respect to the EU budget.

Focusing more centrally on the role of member states, Wurzel *et al.* (2019) also tackle the question of programmatic leadership in another important EU environmental arena, the Council of Ministers and the European Council. Using a leadership framework, they distinguish how the environmental role of the two bodies has incrementally evolved, emphasising that the substantial transformation of enlargement has changed the alliance and leadership dynamics within and between the two bodies. More significant has been institutional change, with new coalition building (including a post-enlargement coalition of East European states that have differed with the leading 'green' EU states on environmental ambition) and an increasing role for the European Council, although only manifesting itself in limited areas such as climate change.

One EU event that many expect to produce a major transformational change in EU integration and environmental policy is Brexit. To assess the environmental ramifications, Burns et al. (2019 – this volume) develop the concept of de-Europeanisation to identify how the UK leaving the EU potentially affects the capacity and resources of actors involved in UK and EU environmental policy. They conclude that, in the short to intermediate term, the path dependent dynamics created by decades of Europeanisation and generally favourable public opinion towards the environment are more likely to lead to drift and limited programmatic change, with major policy change, either in the form of dismantling or innovation, being less likely.

Emphasising the multi-level governance dimensions of the EU and the significant part various territorial actors play in EU environmental policy, Domorenok (2019 – this volume) studies how the design and operational logic of a voluntary policy instrument, the Covenant of Mayors (CoM), created the conditions for potential learning and mutual coordination for local authorities. Examining UK and Italian local authorities, Domorenok finds that, under certain conditions and with often varying motivations, the CoM has gradually empowered local authorities and induced them to play a larger climate change governance role. This effect has, however, been markedly greater in Italy, where the state has been a laggard on climate and energy issues, than in the UK, which has been a leader and which had already charged local government with responsibility for action on energy efficiency.

Focusing more on policy implementation, Börzel and Buzogány (2019 – this volume) examine the roles of member states and the Commission, arguing that the Commission has conducted a successful strategy steering accession states towards compliance and constructing new instruments to achieve this. This contribution also underlines the major environmental policy expansion and wider European integration accomplished through enlargement.

Assessing changes in the enforcement of EU environmental law, Hofmann (2019 – this volume) finds a gradual and significant departure from centralised enforcement, reflecting a more general withdrawal by the Commission from its role as the central enforcer of EU law. The Commission instead supports a move towards improved private, decentralised enforcement of EU law by citizens and nongovernmental organisations. The increasing inclusion of procedural provisions in EU environmental legislation has enabled the EU's individual rights regime to operate in a policy area where it has traditionally been absent; nevertheless, cross-national variation in access to courts and the outlook of civil society actors within each member state delimit the efficacy of this strategy.

This volume highlights important research being carried out on the evolution of environmental policy implementation, but the question of how the EU shapes future environmental policy remains understudied. Schoenefeld and Jordan (2019 – this volume) contend that an important new development in EU environmental policy programme is the gradual rise of *ex post* policy evaluation. They postulate that many motivations may have triggered its gradual growth: they assess the possible role of learning but also argue for examining the notions of accountability and political actors seeking to alter political opportunity structures in their favour.

Conclusions

EU environmental policy has witnessed substantial and significant change since 1992 and especially since 2007. The contributors to this volume suggest that the change has been incremental and rhetorical, despite the major EU changes and crises occurring outside the EU environmental policy sector since 2007. Transformations, such as the enlargements of the 2000s, have seen a continued path dependency of environmental policy and a relatively mixed story concerning implementation, as demonstrated by the decreased Commission ambition (Hofmann) combined with some successful entrepreneurship to boost national capacity (Börzel and Buzogány). New ideas have been generated, legislation such as the Renewables Directive has been approved, soft power continues to be exerted beyond EU borders, and the multiple levels of governance in the EU continue to work on their capacity to contribute to the implementation of EU environmental objectives. Many of these stories have been driven endogenously by dynamics within the policy sector. The impact of Brexit has not altered this picture, certainly in the short term; sustained environmental disequilibria and punctuated equilibria have not happened. This finding is in keeping with those of recent policy dismantling research on the environmental sector (e.g. Gravey and Jordan 2016).

The implications of this finding are more complex than this pronouncement suggests, however. While the findings indicate that the environmental policy sector will continue to be one of the 'success' stories of European integration and held to be such, this picture is less positive for those pushing an environmental agenda, seeking both material and ideational change. Our research suggests a retreat of EU ambition and an inability to consider more demanding alternatives within the EU sector. This may be masked by the creation of new sustainability concepts and statistics about environmental activity and implementation. The contributions suggest that valuation of other policy priorities is higher on the agenda, and this constrains new initiatives and ambitious reformulation. In institutional terms, this is most visibly articulated in the EU Commission's approach to policy formulation and implementation. Exogenous factors such as enlargement and the economic crisis drive processes and relations between actors, with the creation of new alliances and new challenges to the aim of bringing environmental sustainability into the thinking of other EU policy areas.

Comparing this volume with Judge's 1992 Special Issue, we do not see radical differences in either the trajectory of environmental policy or the theories to study it. It is interesting to note that it proved difficult to recruit studies of political parties to link with the study of public policy. While there are researchers working in this field and political party approaches to environmental issues remain important, this reinforces the impression of a policy sector where mainstream (as opposed to green) political parties are not actively positioning themselves with respect to particular policy issues, but where general environmental protection remains a valence issue (Carter 2013). The environmental sector is at a stage where there is a focus on consolidation and improving implementation. The importance of implementation was highlighted in the 1992 issue, but it is significant of contemporary policy developments that much of the present volume's research focuses on implementation and assessment. A related focus different from 1992 is the question of the role of policy instruments, with a number of our contributors highlighting how the selection of instruments and their design have the potential to alter EU politics and policy. There is an ambiguity in this development, with some contributions suggesting that these instruments signify the withdrawal of EU competence while other instruments might indicate expansion.

One dimension that has developed strongly since the 1992 Special Issue is the idea of the EU as a global environmental leader. Particularly in the area of climate change, several of our contributors emphasise the importance of this context for shaping policy choices. Our contributions also show evidence of how climate change has come to dominate also the internal environmental agenda in the past decade. At least one of the contributions (Rietig), however, raises cautions about the trajectory of this leadership. Nevertheless, in an era where the exact leadership role of China and the US is unclear, the EU internal and external policy stances continue to have global significance.

In terms of theoretical perspectives, the contributions do not constitute a radical change from those in Judge's volume. There is a continuing focus on ideas, institutions and alliances. Network analysis did not feature prominently in this volume; discursive analysis features more, but this hardly signifies a trend. One theoretical shift is a greater emphasis on implicit or explicit learning as an explanation of changes, but importantly the authors have tended to combine learning approaches with other perspectives to gain a wider sense of how actors deal with conflicting values and interests, thereby extending our understanding of change and its processes.

In linking the theoretical overview of Table 1 to the findings, one contributor (Hofmann) finds a change in the institutional arena due to international treaty obligations, but generally the contributions focus on changes in ideas, knowledge promoting capacity through instruments and other means, and entrepreneurship. Of the four different eras and policy change over time, the moment of punctuated equilibrium has only arisen with the initial framing of the environmental policy problem in the 1960–1970s. For the rest of the EU environmental policy evolution (including 2007–2017), policy has tended to follow incremental and path dependent lines. This is surprising especially in light of the magnitude of the Eastern enlargement. Spillover was an important dynamic in the 1970–1990s although the changing overall EU policy agenda suggests that this dynamic is now more partial and limited to specific areas than it was before. This incrementalism should not lead us to overlook the fact that significant changes of values have happened over time, with for instance the increasing dominance of neo-liberal thinking (Machin 2019).

Our last reflections concentrate on the broader trajectory of EU environmental policy and EU integration. Our contributors' findings suggest that the EU policy sector has reached a plateau where innovations and ambitious new policy programmes are likely to be limited and constrained for the intermediate term. The state of EU integration suggests that the EU's political agenda will be focused elsewhere. Nevertheless, the brief history of EU environmental policy suggests that environmental problems can effectively seize this agenda and mobilise the public and elites. Such alarms are likely to continue to enlarge the scope of and innovation in environmental policy. In doing so, environmental policy will also remain one of the bedrock policies underpinning EU integration.

Acknowledgments

We thank the two anonymous referees for their comments and the ECPR Environmental Politics Standing Group and ECPR Standing Group on the European Union for their support. Our contribution benefitted from input from participants in the Pisa ECPR workshop, 24–29 April 2016 and 'The Future of Environmental Policy in the European Union Workshop', Gothenburg University,

19–20 January 2017. Anthony Zito acknowledges his British Academy/Leverhulme Small Research Grant 'Towards Smarter Regulation?' SG122203.

Disclosure statement

No potential conflict of interest was reported by the authors.

ORCID

Anthony R. Zito ⓘ http://orcid.org/0000-0002-2312-4781
Charlotte Burns ⓘ http://orcid.org/0000-0001-9944-0417
Andrea Lenschow ⓘ http://orcid.org/0000-0003-2162-1968

References

Baumgartner, F. and Jones, B., 2010. *Agendas and instability in American politics*. Chicago: University of Chicago Press.
Börzel, T. and Buzogány, A., 2019. Compliance with EU environmental law: the iceberg is melting. *Environmental Politics*, 28 (2).
Brady, D., 1978. Critical elections, congressional parties and clusters of policy changes. *British Journal of Political Science*, 8 (1), 79–99. doi:10.1017/S0007123400001228
Brundtland, G. et al., 1987. *Our common future: report of the 1987 world commission on environment and development*. Oslo: United Nations.
Burns, C., et al., 2019. Environmental policy and Brexit: de-Europeanizing or disengaging? *Environmental Politics*, 28 (2).
Burns, C. and Tobin, P., 2016. The impact of the economic crisis on European Union environmental policy. *JCMS: Journal of Common Market Studies*, 54 (6), 1485–1494.
Capano, G., 2009. Understanding policy change as an epistemological and theoretical problem. *Journal of Comparative Policy Analysis*, 11 (1), 7–31. doi:10.1080/13876980802648284
Capano, G. and Howlett, M., 2009. Introduction: the determinants of policy change: advancing the debate. *Journal of Comparative Policy Analysis*, 11 (1), 1–5. doi:10.1080/13876980802648227
Carter, N., 2013. Greening the mainstream: party politics and the environment. *Environmental Politics*, 22 (1), 73–94. doi:10.1080/09644016.2013.755391
Cini, M., 2008. Political leadership in the European commission: the Santer and Prodi commissions, 1995–2005. *In*: J. Hayward, ed. *Leaderless Europe*. Oxford: Oxford University Press, 113–130.
Dealbook, 2007. Paribas freezes funds as subprime woes keep spreading. *The New York Times*, 9 Aug. Available from: https://dealbook.nytimes.com/2007/08/09/bnp-paribas-freezes-funds-on-subprime-woes/ [Accessed 10 June 2017]
Delreux, T. and Happaerts, S., 2016. *Environmental policy and politics in the European Union*. Basingstoke: Palgrave.
Domorenok, E., 2019. Voluntary instruments for ambitious objectives? The experience of the EU Covenant of Mayors. *Environmental Politics*, 28 (2).

Eckstein, H., 1988. A culturalist theory of political change. *American Political Science Review*, 82 (3), 789–804. doi:10.2307/1962491
Edelman, M., 1985. *The symbolic uses of politics*. Champaign, IL: University of Illinois Press.
European Communities, 1987. Single European act. *Official Journal of the European Union L*, 169, 1–28. 29.6.1987.
Flag, B., 2016. Mission and history. http://www.blueflag.global/mission-and-history/, Accessed 31 January 2016.
Gravey, V. and Jordan, A., 2016. Does the European Union have a reverse gear? Policy dismantling in a hyperconsensual polity. *Journal of European Public Policy*, 23 (8), 1180–1198. doi:10.1080/13501763.2016.1186208
Haas, E., 1958. *The uniting of Europe: political, social, and economic forces, 1950–1957*. Stanford: Stanford University Press.
Haigh, N., ed., 2011. *Manual of environmental policy: the EC and Britain*. Harlow: Cartermill.
Hajer, M., 2005. Coalitions, practices, and meaning in environmental politics: from acid rain to BSE. *In*: D. Howarth and J. Torfing, eds. *Discourse theory in European politics*. Basingstoke: Palgrave Macmillan, 297–315.
Hall, P., 1993. Policy paradigms, social learning and the state. *Comparative Politics*, 25 (3), 275–296. doi:10.2307/422246
Hildebrand, P., 1992. The European community's environmental policy, 1957 to '1992': from incidental measures to an international regime? *Environmental Politics*, 1 (4), 13–44. doi:10.1080/09644019208414044
Hofmann, A., 2019. Left to interest groups? On the prospects for enforcing environmental law in the European Union. *Environmental Politics*, 28 (2).
Holzinger, K. and Schimmelfennig, F., 2012. Differentiated integration in the European Union: many concepts, sparse theory, few data. *Journal of European Public Policy*, 19 (2), 292–305. doi:10.1080/13501763.2012.641747
Hooghe, L. and Marks, G., 2009. A postfunctionalist theory of European integration: from permissive consensus to constraining dissensus. *British Journal of Political Science*, 39 (1), 1–23. doi:10.1017/S0007123408000409
Howlett, M. and Cashore, B., 2009. The dependent variable problem in the study of policy change: understanding policy change as a methodological problem. *Journal of Comparative Policy Analysis*, 11 (1), 33–46. doi:10.1080/13876980802648144
Huntington, S., 1971. The change to change: modernization, development, and politics. *Comparative Politics*, 3 (3), 283–322. doi:10.2307/421470
Jachtenfuchs, M., 1996. *International policy-making as a learning process? The European Union and the greenhouse effect*. Aldershot: Avebury.
Jordan, A., 1999. Editorial introduction: the construction of a multilevel environmental governance system. *Environment and Planning C: Government and Policy*, 17 (1), 1–17. doi:10.1068/c170001
Judge, D., 1992. A green dimension for the European community? *Environmental Politics*, 1 (4), 1–9. doi:10.1080/09644019208414043
Kingdon, J., 1984. *Agendas, alternatives, and public policies*. Boston: Little, Brown.
Knill, C. and Lenschow, A., eds, 2002. *Implementing EU environmental policy: new directions and old problems*. Manchester: Manchester University Press.
Knill, C. and Liefferink, D., 2007. *Environmental politics in the European Union: policy-making, implementation and patterns of multi-level governance*. Manchester: Manchester University Press.

Laffan, B., 2016. Europe's union in crisis: tested and contested. *West European Politics*, 39 (5), 915–932. doi:10.1080/01402382.2016.1186387

Latour, B., 1986. The powers of association. *In*: J. Law, ed. *Power, action and belief: a new sociology of knowledge?* London: Routledge and Kegan Paul, 264–280.

Lenschow, A. and Sprungk, C., 2010. The myth of a green Europe. *JCMS: Journal of Common Market Studies*, 48 (1), 133–154.

Lenschow, A. and Zito, A., 1998. Blurring or shifting of policy frames? Institutionalization of the economic-environmental policy linkage in the European Community. *Governance*, 11 (4), 415–441. doi:10.1111/0952-1895.00080

Levy, D. and Newell, P., 2002. Business strategy and international environmental governance: toward a neo-Gramscian synthesis. *Global Environmental Politics*, 2 (4), 84–101. doi:10.1162/152638002320980632

Machin, A., 2019. The more things change, the more they stay the same: the discourse ecological modernisation in the European Union. *Environmental Politics*, 28 (2).

March, J. and Olsen, J., 1989. *Rediscovering institutions*. New York: Free Press.

May, P., 1992. Policy learning and failure. *Journal of Public Policy*, 12 (4), 331–354. doi:10.1017/S0143814X00005602

McCormick, J., 2001. *Environmental policy in the European Union*. Basingstoke: Palgrave.

Moravcsik, A., 1993. Preferences and power in the European community: a liberal intergovernmentalist approach. *JCMS: Journal of Common Market Studies*, 31 (4), 473–524.

Oberthür, S. and Dupont, C., 2011. The council, the European council and international climate policy. *In*: R. Wurzel and J. Connelly, eds. *The European Union as a leader in international climate change politics*. London: Routledge, 74–91.

Pierson, P., 2000. Increasing returns, path dependence, and the study of politics. *American Political Science Review*, 94 (2), 251–267. doi:10.2307/2586011

Real-Dato, J., 2009. Mechanisms of policy change: A proposal for a synthetic explanatory framework. *Journal of Comparative Policy Analysis*, 11 (1), 117–143. doi:10.1080/13876980802648268

Rehbinder, E. and Stewart, R., 1985. *Environmental protection policy, volume 2: integration through law: Europe and the American federal experience*. Berlin: De Gruyter.

Rhodes, R., 1990. Policy networks a British perspective. *Journal of Theoretical Politics*, 2 (3), 293–317. doi:10.1177/0951692890002003003

Rietig, K., 2019. Shifting modes of European climate governance? The European commission's role in reviving climate policy integration in times of multiple economic and security crises. *Environmental Politics*, 28 (2).

Schattschneider, E., 1975. *The semi-sovereign people: a realist's view of democracy in America*. New York: Holt: Rinehart and Winston.

Schmitter, P., 1970. A revised theory of regional integration. *International Organization*, 24 (4), 836–868. doi:10.1017/S0020818300017549

Schoenefeld, J. and Jordan, A., 2019. Environmental policy evaluation in the EU: between learning, accountability and political opportunities? *Environmental Politics*, 28 (2).

Schön, D. and Rein, M., 1994. *Frame reflection: towards the resolution of intractable policy controversies*. New York: Basic.

Steinebach, Y. and Knill, C., 2017. Still an entrepreneur? The changing role of the European Commission in EU environmental policy-making. *Journal of European Public Policy*, 24 (3), 429–446. doi:10.1080/13501763.2016.1149207

Streeck, W. and Thelen, K., 2005. Introduction: institutional change in advanced political economies. *In*: W. Streeck and K. Thelen, eds. *Beyond continuity: institutional change in advanced political economies*. Oxford: Oxford University Press, 1–39.

Thompson, M., Ellis, R., and Wildavsky, A., 1990. *Cultural theory*. Boulder: Westview Press.

True, J., Jones, B., and Baumgartner, F., 1999. Punctuated-equilibrium theory: explaining stability and change in American policymaking. *In*: P. Sabatier, ed. *Theories of the policy process*. Boulder: Westview, 97–115.

Tsebelis, G., 1990. *Nested games: rational choice in comparative politics*. Berkeley: University of California Press.

Van der Straaten, J., 1992. The Dutch national environmental policy plan: to choose or to lose. *Environmental Politics*, 1 (1), 45–71. doi:10.1080/09644019208414008

Vogler, J., 1999. The European Union as an actor in international environmental politics. *Environmental Politics*, 8 (3), 24–48. doi:10.1080/09644019908414478

Weale, A., 1999. European environmental policy by stealth: the dysfunctionality of functionalism? *Environment and Planning C: Government and Policy*, 17 (1), 37–51. doi:10.1068/c170037

Weale, A. and Williams, A., 1992. Between economy and ecology? The single market and the integration of environmental policy. *Environmental Politics*, 1 (4), 45–64. doi:10.1080/09644019208414045

Webber, D., 2014. How likely is it that the European Union will disintegrate? A critical analysis of competing theoretical perspectives. *European Journal of International Relations*, 20 (2), 341–365. doi:10.1177/1354066112461286

Wurzel, R., 2002. *Environmental policy-making in Britain, Germany and the European Union*. Manchester: Manchester University Press.

Wurzel, R., Liefferink, D., and Di Lullo, M., 2019. The council, European council and member states: changing environmental leadership dynamics in the European Union. *Environmental Politics*, 28 (2).

Wurzel, R., Zito, A., and Jordan, A., 2013. *The European government and governance mix: a comparative analysis of the use of new European environmental policy instruments*. Cheltenham: Edward Elgar.

Zito, A., 2000. *Creating environmental policy in the European Union*. Basingstoke: Palgrave.

Zito, A. and Schout, A., 2009. Learning theory reconsidered: EU integration theories and learning. *Journal of European Public Policy*, 16 (8), 1103–1123. doi:10.1080/13501760903332597

Changing the story? The discourse of ecological modernisation in the European Union

Amanda Machin

ABSTRACT
Over the last three decades, ecological modernisation (EM) has emerged as a powerful political discourse, in which economic growth, environmental protection and energy security are mutually reinforcing. Here, the trajectory of EM in the European Union is traced, using a discourse analysis of the seven Environmental Action Programmes. The discourse articulated in these documents points towards an encroaching 'double depoliticisation'. First, political decisions are discursively constructed as a matter of market rationality rather than a democratic process that engages with different political positions. Second, EM is reified as the only feasible solution, and alternative and contesting discourses are marginalised. Thus not only are political differences erased from the discourse, but the discourse is itself removed from political debate.

Introduction

The tale of traditional environmentalism is one of sacrifice and struggle. The heroic figure of the ecologically minded citizen faces the daunting trial of ineradicable environmental limits: sustainability can only be won through the hardship of economic loss. Just as economic profits come with an ecological price tag, environmental goods demand the internalisation of costs by the industries and businesses that have hitherto profited from their externalisation. There is as yet no ending to this story, but it is clear that whatever happens, it is not straightforwardly happy ever after (Maniates and Meyer 2010). In complete contrast, however, a distinctly modern account has arisen, which inverts the apparent truism of the traditional tale. 'Ecological Modernisation' (EM) offers storylines in which economic and environmental goals are no longer pitted against each other, but rather are neatly reconcilable. As Weale and Williams explain, according to EM: 'environmental protection should be seen not as being in competition with economic growth and development but instead as an essential precondition for such growth and development' (1992, 47). The storylines of EM herald a much happier ending, and thus it is no surprise that it increasingly dominates within political rhetoric and frames policy-making at various levels of governance.

The European Union (EU) has long been heralded as a global leader of environmental governance (Zito et al. 2019 – this volume) and therefore EU environmental policy is a particularly appropriate arena to assess the trajectory of EM over the last three decades. To do so, I begin by offering an account of EM as a powerful political discourse that has become prominent across a broad spectrum of academic theory and policy practice. I then consider the emergence of the EM discourse in the EU by analysing the Environmental Action Programmes (EAPs) – important documents that have informed and reflected the goals and trajectory of EU environmental policy. My findings suggest that despite the economic crisis of 2008, which we might have expected to disrupt the discourse, or at least provoke a reassessment, EU environmental policy strategy has actually reaffirmed EM. The dominance of EM within the EAPs is also arguably reflected by the emergence of the Emissions Trading System (EU-ETS) as the key policy instrument underpinning flagship climate policies, an instrument in which market forces are supposedly harnessed for environmental ends.

My central claim is that the rising dominance of this discourse points towards a tendency of *double depoliticisation*. Within the EM discourse, political dissent is smoothed over by economic rationality; market competition and innovation replaces political regulation. Further, the discourse itself is reified as the only feasible strategy, a matter of 'common sense' and therefore one that is 'outside' or 'beyond' politics. Not only is politics taken out of the discourse, but the discourse is taken out of politics. This 'double depoliticisation' is, of course, highly political. If we accept that environmental policy-making is not a matter of implementing and managing predetermined options, but ultimately a matter of making decisions with which not everyone will concur, and which could always be otherwise, then environmental policy-making is deeply and inevitably political (Remling 2018, 478). Any discourse that seems 'natural' is concealing the 'undecidability' of the terrain of politics (Laclau and Mouffe 2001, xi). What is at stake here is the extent to which EM has become a hegemonic political discourse, by which I mean the degree to which its powerful storylines have become reified and are therefore difficult to challenge by those who might offer alternative perspectives, visions and agendas. Although I do not attempt to pinpoint the causes of this, various interconnecting factors have led to the general acceptance and normalisation of the EM discourse, not only by EU policy-makers but across the various levels and arenas of environmental governance.

The weakness of EM emerges, however, precisely because of its apparent strength: the failure of universal green benefits to take root undermines EM's construal of a 'happy ending' in which all benefit from a 'greening of the economy'. The feasibility of the storyline varies between and within different states (Curran 2009, 205). Alternatives – such as a 'limits to growth' or 'post growth' discourse – are, however, rendered less visible and regarded as unrealistic. As other recent analysis in this journal observes, the depoliticisation of environmental policy that is occurring at the EU level protects the status quo (Remling 2018). Here, I offer an empirically grounded critique of a highly problematic depoliticising political discourse dominating the EAPs of the EU.

The rise of EM

EM has emerged over the last three decades as a familiar term to frame and inform environmental policy-making at various levels of governance, particularly in the Western industrialised states. Nevertheless, and perhaps not unrelated to its wide usage, its meaning remains indistinct and varied (Buttel 2000, 59, Christoff 1996). It is used as: a *descriptive* tool to analyse the dynamics of environmental reform; a *normative* term to promote certain socio-economic approaches (Mol and Sonnenfeld 2000, 7, Bailey *et al.* 2010); a *strategy* of using technical innovation to make environmental improvements (Baker 2007; Jaenicke 2004); a *policy paradigm* (Toke 2001; Szarka 2012); a *category* containing *any* environmental improvement (Buttel 2000, 60); and a socio-political *discourse* (Hajer 1997).

I follow Hajer in understanding EM as a *policy discourse* in which environmental protection makes economic sense (1997, 3). For Hajer, in line with the work of Foucault, a discourse is not simply a unit of written or spoken communication as some academic disciplines hold (Hewitt 2009, 2). Rather, it is 'an ensemble of ideas, concepts and categories through which meaning is given to phenomena, and which is produced and reproduced through an identifiable set of practices' (Hajer 2005, 303). Depicting EM as a discourse that generates meaning allows us to understand that EM is not simply a rational response to environmental 'facts', but rather a way of constructing those facts in the first place. Discourses present 'succinct and agreeable' storylines that incorporate diverse actors and elements (Jessup 2010, 22). When a discourse becomes dominant, its particular storylines become 'common sense' (Hajer 2005, 3203).

The EM discourse arose in response to prominent 'de-modernisation' arguments in the 1980s, which asserted that tackling environmental problems demanded a fundamental rethinking of modern ways of life (Mol and Spaargaren 2000, 19). The famous 1972 report, *The Limits to Growth*, had emphasised the existence of limits to industrial and social expansion (Meadows *et al.* 1972). EM offered a timely counter claim, portraying economic growth and environmental protection as not antagonistic, but rather mutually reinforcing: market dynamics are regarded 'as carriers of ecological restructuring and reform' (Mol and Sonnenfeld 2000, 6). EM attempts to solve ecological crisis through economic growth (Baker 2007, 304). Not only does the discourse suggest that the environment might benefit through the innovation encouraged by market mechanisms, it also asserts that, conversely, the market can benefit from environmental challenges by implementing strategies and tools to protect the environment, industries and businesses enhance their efficiency and the rising demand for green technology drives innovation and development (Baker 2007, 299). In short, EM promotes the idea that 'pollution prevention pays' (Dryzek 2005, 167); early action forestalls expensive clean-up operations and a cleaner healthier environment means happier and more productive employees (Dryzek 2005, 168). EM articulates ecological risk not as a *limit* to growth but rather as a *catalyst* for change and can be understood as both product and critique of modernity (Christoff 1996).

More recently, the theme of energy security made an appearance within EM too. Germany's domestic environmental policy, for example, connects sustainability not only to economic gain but also to enhanced security: a shift to greener energy sources reduces reliance upon supplies of natural gas coming from Russia. Thus 'the energy security debate reinforces Germany's EM agenda' (Hillebrand 2013, 677). By the early 1990s, EM had become a dominant policy discourse across Europe and beyond (Buttel 2000), articulating a storyline of an apparent rational and inclusive consensus in which all economic sectors, investors, entrepreneurs, public authorities, citizens as well as the environment stand to gain (Curran 2009, 203).

Expectations regarding the role of government regulation remain ambiguous; in some of the storylines of EM, the state plays a continued, albeit transforming, role alongside other political institutions, non-state actors and social movements in implementing policies to guide the market towards ecological goals (Mol and Sonnenfeld 2000, 7, Dryzek 2005). Stringent governance encouraged the necessary transformations and innovations (Berger *et al.* 2001, 57). The assumption here is that while capitalism can be greened, and economic competition can be harnessed to make production and consumption more sustainable, this can only occur '*under certain political conditions*' (Buttel 2000, 61 – emphasis added). This is why scholars have acknowledged that structural change is often needed to promote EM, or to step in where EM is not sufficient (Jaenicke 2004, 204).

This element of governance is not always emphasised; in line with a general shift from centralised government to decentralised 'governance', EM articulates market mechanisms, producers and consumers as the agents of reform (Berger *et al.* 2001, 59). These economic agents, however, are discursively constructed as motivated by economic rationality, not environmental concerns (Berger et al. 2001, 60). Already in 2007 Baker observed a reliance, in the EU at least, upon 'light governance tools' in which state intervention is minimal; it is simply taken for granted by policy-makers that businesses will voluntarily and independently pursue a strategy towards sustainability (2007, 309). EM constructs markets as capable of addressing climate change with only 'minor tinkering' (see Glasson 2012). As Christoff has noticed, in so far as EM does not challenge traditional capitalist imperatives, it may simply put a 'green gloss' on industrial development (1996, 486). The discourse merely reaffirms the trust in industrial progress and therefore: 'there is a danger that the term may serve to legitimise the continuing instrumental domination and destruction of the environment' (1996, 497). Christoff's hope for a 'stronger' and more ecological EM notwithstanding, in its 'weak' forms, it places exclusive emphasis on technology, entrepreneurs and market instruments underpinned by an orientation towards innovation and profit, rather than social, economic and lifestyle change and strict state regulation by the state (Baker 2007, 303–4, Bailey et al. 2010, 5). So, although there is clearly politics at work in implementing the strategies of EM, this is frequently underplayed.

Theorists have long argued that a crucial weakness of EM is that, by reaffirming the centrality and logic of economic growth, EM is unlikely to lead to a decrease of *total* resource consumption (Jaenicke 2004; Bailey et al. 2010). A more radical shift to sustainability would instigate structural transformation and obstruct the drive towards short-term economic profit (Pepper 1998). Critics therefore warn that 'the pace of global environmental change is out of sync with the pace of institutional reform advocated by EM' (Warner 2010, 553). Here, I do not dissent from these concerns, but I focus upon a different issue. I consider how the changing discourse of EM indicates a trend to what I call a 'double depoliticisation'.

EM in the EU

The EU approach to environmental policy-making is extremely influential. As has been well documented, the EU has emerged as a leader in environmental politics in general and in climate change politics in particular (Oberthuer & Kelly 2008; Skovgaard 2014; Baker 2007; Fischer and Geden 2015). Climate change offers a particularly 'intractable policy problem' (Zito et al. 2019), which potentially challenges traditional policy-making processes and institutions. The EU seemingly initially rose to this challenge: in 2008, EU legislation enshrined a commitment of cutting greenhouse gas emissions by 20% by 2020 compared to 1990 levels (Skovgaard 2014). In 2011, the European Commission presented a longer term strategy in the 'Energy Roadmap 2050' which expresses the aim of cutting domestic emissions by 80–95% by 2050 compared to 1990 (European Commission 2012).

The EU's commitment to environmental goals allows it to differentiate itself from other actors and to forge a common identity (Baker 2007, 312, Oberthuer and Roche Kelly 2008; Fischer and Geden 2015). Since pollution problems tend to be transboundary, it is clear why environmental policy would play a part in European integration (Weale *et al.* 2000, 29). As Zito *et al.* (2019) observe in their introduction, the imperatives of the single market programme, regarded as the central instrument of European integration, inevitably had an effect upon EU environmental policy. Policy-makers were aware that national environmental legislation could impede the functioning of the market and create barriers to free trade (Weale *et al.* 2000, 30–32). Yet this also precipitated the idea that environmental policy could instead support the emergence of the single market and the globalisation of trade, and that it was precisely through high standards that the EU could become competitive against the USA and Japan (Weale *et al.* 2000, 37). In the 1970s and early 1980s, environmental policy was focused upon regulation. But by the late 1980s and early 1990s, European policy elites were articulating the idea that the imposition of high environmental standards did not have to contradict the goals of the single market (Weale *et al.* 2000, 78). I argue, in agreement with Baker (2007), Weale and Williams (1992) and Weale *et al.* (2000), that EM has undeniably been a crucial component of the EU's environmental policy, and that it continues to be so. Thus, the Lisbon strategy of 2000 adopted a vision of 'smart sustainable and inclusive growth'. The *Energy Roadmap 2050* reaffirmed this goal: 'continued progress is needed to a low-carbon economy with new opportunities for growth and jobs' (European Commission 2015, 2).

It may well be that policy-makers have continued to find EM appealing, particularly after the economic crisis, *precisely because* it permitted continued emphasis on environmental goals regardless of any perceived contradiction to prioritising economic issues. The introduction to *the Energy Roadmap 2015* explicitly presents this argument:

> Today, public deficits, jobs and pensions seem more important than future energy needs. Yet by investing in our energy system, we create jobs, businesses and prosperity. Less energy wastage and lower fossil fuel imports strengthen our economy. Early action saves money later. (European Commission 2012, 1)

Likewise, the Commission has emphasised the compatibility of environmental law with an agenda for growth (Hofmann 2019 – this volume).

To be clear: I am not implying that EU policy-makers were necessarily consciously utilising a deliberate rhetorical strategy, but rather that the discourse has powerfully constructed a social reality that was attractive, useful and therefore accepted, rearticulated and sedimented as 'commonsense'. It is important to ask, however, whether this strategy works in favour of some more than others and whether it works at all in favour of the environment. I suggest that EM has become so hegemonic that it limits the possibility of asking these questions. This is what I mean by 'depoliticisation' – the removal of the possibility of political dissent. EM is presupposed and promoted as the only feasible option.

In the next section, I map the trajectory of the EM discourse, alongside other discourses, in the EU over the last two decades by analysing the seven EAPs.

EAPs

The EAPs of the EU are key documents for EU environmental policy. Composed by the Commission, the EAPs are medium-term plans that guide, frame and chart its overarching and evolving priorities, principles and programmes and therefore 'set the course for forthcoming initiatives and legislative proposals' (Endl and Berger 2014, 22; see also Baker 2007, 304). These documents not only frame and inform policy-making, but they encapsulate and reaffirm the dominant discourses, reflecting some of the key aspects of environmental policy-making (Hey 2006, 18).

The seven EAPs cover, in succession, the period from 1973 up to 2020. They provide a rich historical record of the changing EU environmental policy and its guiding principles and dominant discourses. Prior research has suggested the prominence of EM discourse in the EU (Baker 2007; Weale and Williams 1992). Here, I aim to add to this scholarship by mapping its trajectory over time by considering whether and in what ways the EAPs resonate with this discourse. David Judge notes that these documents are multifaceted and can be read in different ways (1992, 8). They are what Foucault would term socio-political *artefacts*, which can be interpreted not as straightforward communication from policy-makers, but as revelatory of underlying ideological assumptions, contradictions and lacunae.

The aim of deconstructing policy texts is not to uncover the author's intended meaning but to discover their effect on the reader. What is crucially important is not *what is meant* but rather *what is actually written* (Foucault 1991, 63). As Cod observes: 'the policy text, in spite of itself, embodies incoherences, distortions, structured omissions and negations' (1988, 245). Yet such texts reveal rules regarding what can and cannot be said. Foucault asks: 'which utterances can be put into circulation... Which are repressed... Which utterances does everyone recognise as valid, or debatable, or definitely invalid? Which have been abandoned as negligible, and which have been excluded as foreign?' (1991, 60).

There are many variations of discourse analysis. I use the Foucauldian variation presented by Hajer (1997, 2005), since he offers a seminal account of EM as a policy discourse. In undertaking my analysis, I focused on two particular features of a discourse: 'storylines' and 'subject positions'. Hajer uses the term 'storyline' to refer to a 'condensed statement' that simplifies a complex topic (2005, 302). As he explains: 'political change may well take place through the emergence of new storylines that re-order understandings' (Hajer 1997, 56). By 'subject positions', I refer to the locations or identities made available for a subject in a discourse structure (Laclau and Mouffe 2001, 115). Foucault, along with other theorists of discourse, challenges the idea of a 'sovereign subject' manipulating passive discourses. Rather, certain discourses enable and constrain different forms of subjectivity that are 'positioned' within the discourse (Hajer 1997, 48). The dominant storylines in a discourse promote and demote particular subject positions. Therefore, I am not only interested in the words that are printed in the EAPs, but in the storylines understood as 'common sense' and the empowerment of certain subjectivities; I analyse how discourses organise meaning in the EAPs.

I started by taking each EAP as a unit of analysis. The initial stage was a close *comparative* reading of these units with the aim of noticing distinct similarities and differences. My aim was to discover dominating storylines and subject positions that both *appear and disappear*. In this first reading, I identified prominent subject positions identified and then, with the software MAXQDA, subjected them to a frequency count across each unit (see Table 1) taking care to only include relevant instances. However, the *frequency* of a term or phrase does not necessarily correlate with its *empowerment*. A subject position may be articulated as a passive audience or an active participant. I conducted a second reading in order to determine who and what was given priority and agency within the text and thus who and what might be empowered within and by the discourse. I identified

Table 1. Frequency count of 'subject positions' in European Union (EU) Environmental Action Programmes (EAPs).

	Page extent	Expert	Consumer*	The public**	Citizen	Tourist***	Business	Stake-holder
EAP1	54	29	4	2	0	1	0	0
EAP2	47	18	5	7	0	1	0	0
EAP3	25	2	1	6	0	0	0	0
EAP4	45	1	9	17	3	1	0	0
EAP5	93	6	40	38	22	22	10	0
EAP6	81[a]	3	23	19	28	4	31	11
EAP7	92[a]	1	8	20	27	0	22	9

[a] Note that unlike previous EAPs, 6 and 7 were published in a different format and include pictures.
* **Includes** 'consumer demand'
** **Includes** reference to 'public' as 'public sphere': General Public/The Public/Public Opinion/Public Awareness/Public Interest/Public Demand/Public Information/Public Participation/Public Action/Public Concern/Public Dialogue(s)
 Excludes reference to 'public' as in 'state': Public Authority(ies)/Public Sector/Public Transport/Public Goods/Public Revenue/Public Procurement/Public–Private Initiatives/Public Body(ies)
*** **Excludes** 'tourist industry'/'tourist sector'/'tourist trade'

storylines by recording prominent assertions regarding the way the problems, contradictions and complications in environmental politics are handled.

In his summary of the first six EAPs, in which he documents their specific policy approaches as well as resistance to them, Hey suggests that they are characterised more by continuity than by change (2006, 19). I examine this claim and conclude that there are several shifts in discourse, but nevertheless a persistent trajectory towards a 'doubly depoliticised' EM that is most unambiguously articulated in EAP7. Table 2 presents a summary of my findings.

EAP1 (1973-1976)

As Hey notices, the first two EAPs emphasise the need for a comprehensive assessment of the environmental challenges (2006, 18-19). Indeed, EAP1 asserts above all the importance of 'objective analysis of the facts' (European Commission 1973, 8). It acknowledges the critical role played by environmental policy for quality of life for the peoples of the community, ensuring 'sound management' and the finding of 'common solutions' to 'prevent, reduce and as far as possible eliminate pollution and nuisances' and to not only *protect*, but to *manage* and also *improve* the environment. It emphasises the importance of education to bring awareness and acceptance of citizens 'responsibility towards the environment' (pp. 11 and 46-47). EAP1 asserts the 'polluter pays principle' but suggests that its implementation must be worked out to 'avoid distortion of trade and investment' (p. 9). On the one hand, it states: 'economic expansion is not an end in itself' (p. 5). On the other, it is also clearly stipulated that environmental policy 'should be carried out in such a way as does not jeopardise the satisfactory operation of the common market' (p. 8). While there is no acknowledgement of any *conflict* between environmental and economic goals, there is also no acknowledgement of any potential mutual benefit. Rather they must be *reconciled*: 'environment policy can and must be compatible with economic and social development' (p. 6). The EU should adopt measures 'for the protection of the environmental by reconciling that objective with the satisfactory operation of the common market' (p. 9).

EAP2 (1977-1981)

EAP2 re-emphasises the importance of the improvement of scientific and technological knowledge and 'technical progress' (European Commission 1977, 7). It also re-emphasises the importance of international cooperation. However, in contrast to EAP1 it strikingly stresses the *natural limits* to economic growth. 'Nature pays a considerable price for economic

Table 2. Summary of discourse analysis of EAPs 1–7.

Period	Dominant storylines	Empowered subject positions	Key discourse(s)
EAP1 1973–1976	Greater research determines 'the facts' and standardises measurement of pollutants. Education will teach population to accept environmental responsibility	The expert	*Knowledge speaks truth to power*
EAP2 1977–1981	There are natural limits to resources and to economic activities. Environmental policy constrains economic development but does not conflict with it	The expert	*Knowledge speaks truth to power and limits to growth*
EAP3 1982–1986	Environment provides both limits to and possibilities for economic activity. Environmental policy can stimulate employment and technological innovation, to aid economic competitiveness	The public	*Limits to growth and ecological modernisation*
EAP4 1987–1992	Environmental policy is an economic opportunity. Economic instruments can be used to ensure that 'the polluter pays'	The public, the consumer	*Ecological modernisation*
EAP5 1993–2000	Economic growth can continue as long as it respects environmental conditions. The shift to sustainability enhances competitiveness and innovation	The consumer, the public, the tourist	*Sustainable development and ecological modernisation*
EAP6 2002–2012	Economic growth and environmental damage can be 'decoupled'. Sustainable development offers business opportunities	The citizen, the business, the consumer	*Ecological modernisation and sustainable development*
EAP7 2012–2020	The 'Green Economy' can absolutely decouple economic growth and environmental degradation	The business, the citizen, the stakeholder	*Ecological modernisation*

expansion and as a result, some production possibilities are reduced while their production costs... are sometimes increased... material growth has physical limits' (p. 5). It therefore states that one of its objectives is to 'guide development in accordance with quality requirements' (p. 6). EAP presents environmental policy as placing constraints on economic activity to introduce 'reasonable and consistent' structural changes (p. 37). Somewhat contradictorily, however, it also explicitly states that: 'an environmental policy *does not conflict with economic development*. Indeed the lack of such a policy may in itself place constraints on the development of some economic activities' (p. 36). It also explains that: 'the implementation of environmental measures will generally encourage industry to perfect less expensive anti-pollution techniques' (p. 37).

EAP3 (1982-1986)

There is a fundamental shift in the third EAP. The notions of environmental *limits* and of environmental *protection*, dominant in EAP2, are overshadowed in EAP3 by notions of *opportunities* afforded by environmental policy. The text on the very first page states: 'account should be taken of the economic and social aspects of environmental policy, and particularly of its potential to contribute to the easing of current economic problems, including unemployment' (European Commission 1983, 1). The document no longer understands environmental protection measures to be necessary primarily for improving quality of life, but rather to 'support and complement economic development' (p. 3) and to 'find solutions' to economic problems (p. 4). The potential of the creation of jobs, not mentioned by the previous two programmes, now features on several occasions (p. 4) as does technological innovation (pp. 5 and 7). As Hey notes, in EAP3 and EAP4 there exists 'an assumption of harmony between the objectives of the internal market and environmental protection' (2006, 20).

In an early paragraph in both EAP1 and EAP2, it was stated that among the fundamental tasks of the community were 'the improvement of the quality of life and protection of the natural environment' (European Commission 1973, 2, 1987, 1). In EAP3, the authors have adjusted this text to state that among the fundamental tasks are 'the improvement of the quality of life and *making the most economical use possible for the natural resources* offered by the environment' (European Commission 1983, 1 - emphasis added). The natural environment no longer is in need of *protecting*; while it is construed as a constraint (p. 4), it is also constructed as an *economic resource* (p. 3). There is, certainly, a strengthened emphasis upon the importance of environmental policy and the integration of the environmental dimension into other policy areas (p. 2), but this is also articulated as a way to strengthen economic competitiveness. In short, while EAP3 does not completely drop the notion of *environmental constraints*, it is now complemented by the concept of *economic possibility*.

EAP4 (1987-1992)

The fourth EAP consolidates the shift to EM. Here it is concisely and baldly stated that: 'the protection of the environment can help to improve economic growth and facilitate job creation' (European Commission 1987, 2 and 7). Environmental protection has become an 'economic imperative' (p. 6) and 'an essential element in the future economic success of the Community' (p. 7). Furthermore, this challenge is explicitly regarded as an opportunity (p. 7). The notion that *the polluter should pay*, mentioned in previous EAPs becomes a fully fledged principle in EAP4. In order to ensure 'the polluter pays' EAP4 discusses both legal regulations as well as economic instruments (p. 15). It expresses the intention to use economic instruments in various policy fields (air pollution, water pollution, protection against noise, nature protection and waste management) (p. 15).

The idea that the population must simply be informed about their environmental responsibilities has faded. Where the text discusses education, the emphasis is on rendering the policy-making process transparent and shoring up 'public acceptance' (p. 15). It is only children who need educating. Instead of a top-down dissemination of information to the population, the document constructs public as a subject-position whose opinion, attitude and acceptance should be taken into account: 'Much more could be done to inform the public and thus influence public opinion in favour of strict environmental policies' (p. 16). Technical reports should thus be 'published in appropriate ways' in order to make information accessible to all (p. 16).

EAP5 (1993-2000)

There is another significant shift in the fifth EAP. Entitled 'Towards Sustainability', EAP5 articulates the *Sustainable Development* (SD) discourse and explicitly references the *Brundtland Report* (1993, 12). According to the storyline of the SD discourse, socio-economic development can, and should, continue, but in a way that does not undermine the very environmental conditions it relies upon. SD problematises the traditional growth paradigm (Baker 2007, 303) and 'challenges the industrialised world to keep consumption patterns within the bounds of the ecologically possible' (Baker 2007, 302). The EAP reiterates this idea: 'the ultimate limiting factor for continued efficiency and growth as they interface with one another is the tolerance level of the natural environment' (European Commission 1993, 24). Here, then, fundamental limits are articulated, yet rather than a fixed cut-off point, as appears in the storyline of the 'Limits to Growth' discourse, these limits provide ongoing boundary conditions for economic development. Thus, the intention is to promote continued economic and social development without causing environmental harm (p. 12). The EAP5 goes as far as making the strong statement that: 'The real problems, which cause environmental loss and damage, are the current patterns of human consumption and behaviour' (p. 13).

In ambiguous tension with this SD storyline, however, is another less explicit EM one, that the requirements for staying within these boundary conditions will have economic *benefits*. For example: the introduction to EAP5 contains this classic statement of SD: 'the increased economic growth expected will be unsustainable unless environmental considerations are taken into account...' But the sentence then continues: '...not so much as a potential limiting factor, but rather as an incentive to greater efficiency and competitivity with particular reference to the wider international market-place' (p. 20). Here, sustainability is not construed as *constraining* economic activity but rather as *stimulating* it. 'Rather than reduce competitive advantage, stringent environmental requirements can actually enhance it by triggering upgrading and innovation' (p. 31). As Baker notes, these two discourses are sharply contrasting and ultimately incompatible (2007, 301–304).

Another tension exists between the role of government and the market. It is clearly stated that government should not only provide legislation, but also in 'creating the necessary framework and conditions' to allow 'the individual and private enterprise to play their respective role' (European Commission 1993, 26). Indeed, EAP5 refers to individuals both as citizens and as consumers, who are, of course, market participants. Information is needed to be made available to the public, not only to develop environmental awareness (p. 14) and public involvement (p. 26) but to improve 'consumer choice' (pp. 27 and 29).

EAP6 (2000–2012)

The sixth EAP ostensibly continues the SD discourse (European Commission 2000, pp. 4 and 6), promoting a 'decoupling' of economic growth and environmental degradation (p. 9). The environment 'must be taken seriously' (p. 5). However, environmental protection – and SD itself – is not only an imperative, but also an *opportunity*: 'High environmental standards are also an engine for innovation – creating new markets and business opportunities' (p. 9). 'If tackled in the right way, our efforts to limit climate change are likely to generate significant opportunities and benefits for business as well as side benefits it terms of reduced air pollution' (p. 29).

EAP7 (2012–2020)

A major storyline in the EM discourse is that of the 'green economy' (ESDN 2014, 44) in which there is 'an absolute decoupling of economic growth and environmental degradation' (European Commission 2014, 7). Not only can growth still occur, but focus upon the environment will bring *new* sources of growth and jobs (p. 8). The document clearly articulates the EM construal of a 'win-win' situation:

> The 7th EAP reflects the Union's commitment to transforming itself into an inclusive green economy that secures growth and development, safeguards human health and well-being, provides decent jobs, reduces inequalities and invests in, and preserves biodiversity, including the ecosystem services it provides (natural capital), for its intrinsic value and for its essential contribution to human well-being and economic prosperity. (p. 19)

The EAP sets the overall objective of the EU becoming 'a smart, sustainable and inclusive economy by 2020 with a set of policies and actions aimed at making it a low-carbon and resource efficient economy' (p. 4). This overall objective is formulated into nine priority objectives (PO). Most interesting for my analysis is PO2: 'To turn the Union into a resource-efficient, green and competitive low-carbon economy'. Here, EM is clearly reaffirmed and neatly encapsulated. Importantly, government action becomes a matter of providing 'the right framework conditions for investment and eco-innovation, stimulating the development of sustainable business or technological solutions' (p. 33). There is a shift away from legislation towards market-based instruments.

Notably, 'stakeholder' now appears more frequently than 'consumers'. Although it appears less frequently overall than 'public', EAP7 articulates it in a different way; the public is usually a passive subject-position; a member of the public in this document is someone who is concerned, and needs reassurance, information and well-being. In contrast, the 'stakeholder' is sought after as someone who is interested and should be 'engaged' (p. 9) 'involved' (p. 75) and 'consulted' (p. 11) as an active participant in decision-making – for example regarding sustainable urban development (p. 76). However, the identity of these 'stakeholders' is left unqualified. If a stakeholder is someone who holds 'a stake' then are they not *already* engaged? What about those who are not seen, who do not see themselves, as holding a stake? This is not addressed in EAP7. Such vagueness might justify the suspicion that the statement of intention to consult 'interested stakeholders' may work as little more than a rhetorical device to connote inclusive and democratic political decision-making without actually explaining how this can work in practice.

The dominance of EM

Hajer explains that a discourse becomes dominant when 'central actors [are] forced to accept the rhetorical power of a new discourse' ('discourse structuration') (2005, 302) and when it 'solidifies in particular institutional arrangements' ('discourse institutionalisation') (2005, 303). I suggest there is evidence of both 'structuration' and 'institutionalization' of the EM discourse in the environmental policy of the EU.

With regards to *structuration*, the analysis of the EAPs reveals the way in which EM has become a hegemonic discourse. It is possible to see the *institutionalisation* of EM in the EU's ETS. EM makes market mechanisms 'common-sense' policy instruments. Thus, as various commentators have observed, environmental policy discussions frequently weigh the merits of emission taxes (Meckling and Jenner 2016) against those of emissions trading (Stephan and Paterson 2012), these two instruments have been accused of 'crowding out' other options (Stavins 2008). Indeed, while other policy options and instruments exist, as the largest emissions market in the world, the EU-ETS is regarded as a 'cornerstone' in the EU's climate policy (Convery 2009; Bailey 2010; European Commission 2016). As EAP7 states, 'The Union Emissions Trading System will continue to be a central pillar of Union climate policy beyond 2020' (European Commission 2014, 35). The ETS fits within EM, but EM also features within the ETS, which the EU-ETS handbook claims 'contributes to the creation of jobs, generation of green growth and strengthening long term competitiveness of the European economy by putting a price on carbon' (European Commission 2015, 14). Important problems that have hampered the effectiveness of this policy instrument (Schuppert 2011; Bailey *et al.* 2010, 10) have not yet been resolved, but what is particularly significant for my argument, is that the focus upon emissions trading has reduced the space for the emergence of alternatives and suppressed the expression of more radical voices, schemes and policies (Bailey *et al.* 2010, 15).

It is possible to see, then, the further 'double depoliticisation' of EM in the institution of the EU-ETS. First, although governance mechanisms set up and regulated the ETS, there is a heavy deference to the market; its 'fact sheet' lionises the harnessing of market forces to find the most cost effective way of reducing emissions and giving companies flexibility (European Commission 2016, 2). The decisions that are taken regarding this instrument work mainly in the interests of industry with a focus of provoking innovation and incentivising energy efficiency. The growing prioritisation of the EU-ETS as a central policy instrument illustrates a reliance on market rationality to implement the necessary structural behaviour change, reducing the need for any explicit political decision or governmental intervention (Bailey *et al.* 2010, 5). Second, any dissonance regarding the role of the ETS as 'the central pillar of EU climate policy' that '*will have to* play an increased role' (European Commission 2012, 16 – emphasis added) is simply not acknowledged.

This is not to say that regulation plays an insignificant role in EU environmental policy. But the important policy instruments that work through regulation rather than market innovation are seemingly often portrayed, in line with EM, as 'good' for business. For example, the 2006 regulation over the EU chemicals industry (REACH): 'should ensure a high level of protection of human health and the environment... *while enhancing competitiveness and innovation*' (EU 2006, 2 – emphasis added).

Conclusion

While earlier economic recessions have arguably contributed to shifts in EU environmental policy (Ward 1997, 179), the severe impacts of the more recent economic crisis did little to provoke a change in discourse; as Zito *et al.* (2019) document, no 'punctuated equilibria' or 'sustained disequilibria' have occurred. Some expected environmental issues to be placed on the 'political backburner' but others held that economic and environmental challenges could be tackled together through 'green growth' (Tienhaara 2014, 187). Since 2008, environmental concerns have been articulated by various actors in politics and the media as a market opportunity (Tienhaara 2014, 188). The EM policy discourse thus remains intact, continuing to dominate EU policy documents and instruments. This is hardly surprising. The storyline of the EM discourse reconciles economic growth and ecological targets to offer a 'win-win' situation. It offers an apparently perfect solution at a time when the stakes are rising over economic and environmental decisions alike.

One crucial question here is whether the depoliticised assertion of EM as rational, realistic, consensual and beneficial to all disguises the fact that, in contrast to its articulation of a win–win situation, the discourse of EM suits some more than others. The goal of EAP7 is to turn the EU into a resource-efficient, green and competitive low-carbon economy – but the viability of this goal is highly dependent on the present situation in a given country (Jordan & Lenschow 2000, Lekakis and Kousis 2013). Are national economies equally capable of gaining economically from renewable energy production? Are they equally able to provide 'green jobs' and remain competitive? And how much do the prospects of economic growth stemming from green technologies vary? In articulating a storyline where everyone wins, these political questions are prohibited by EM discourse. So are challenges regarding its effectiveness in achieving environmental goals. Politics is effaced from the discourse: on the one hand, politics is rendered unnecessary, since market mechanisms can deliver the solution: on the other, politics is made impossible, because the discourse is 'common-sense'.

This trend towards 'double depoliticisation', indicated and instigated by the highly political EM discourse is extremely problematic, for it disguises the disagreements at play in environmental politics and forestalls the emergence of alternative approaches (Machin 2013, 16–24). But might these alternatives offer more radical ecological *and* more progressive economic strategies and policies? Might it be possible to open up policy-making to a more 'agonistic' contestation between alternative and conflicting discourses and strategies? Any real shift to sustainability surely must be a result of political decisions that do not, and cannot, satisfy everyone and everything; stricter and more explicit regulation and a brake on capitalist expansion may well be both ultimately requisite and hotly contested. But the hegemonic EM discourse precludes the expression of differences across all levels and arenas of political discussion, instead its depoliticised deference to the market risks producing a 'hollowed out' democracy and declining trust in political representatives (Ruser 2015). By disavowing radical transformation, such a depoliticised approach offers and demands 'more of the same' (Remling 2018, 491).

The EM storyline might be appealing, especially in the short term. But its happy ending seems unlikely, forestalled by the political disagreements that it refuses to acknowledge. It is possible that the discourse of EM embrangles the very problem it tries to resolve.

Acknowledgments

I gratefully acknowledge helpful comments on previous drafts made by Anthony Zito, Charlotte Burns, Andrea Lenschow and Charlotta Söderberg. I also benefitted immeasurably from discussions at the 2016 ECPR Joint Sessions in Pisa and 'The Future of Environmental Policy in the EU' Workshop in Gothenburg organised by Andreas Hofmann.

Disclosure statement

No potential conflict of interest was reported by the author.

References

Bailey, I., 2010. The EU emissions trading scheme. *WIREs Climate Change*, 1 (1), 144–153. doi:10.1002/wcc.17

Bailey, I., Gouldson, A., and Newell, P., 2010. Ecological modernisation and the governance of carbon: a critical analysis. *The Governance of Clean Development: A Working Paper Series*. Centre for Climate Change Economics and Policy.

Baker, S., 2007. Sustainable development as commitment: declaratory politics and the seductive appeal of ecological modernisation in the European Union. *Environmental Politics*, 16 (2), 297–317. doi:10.1080/09644010701211874

Berger, G., et al., 2001. Ecological modernization as a basis for environmental policy: current environmental discourse and policy and the implications of environmental supply chain management. *Innovation*, 14 (1), 55–72.

Buttel, F.H., 2000. Ecological modernization as social theory. *Geoforum; Journal of Physical, Human, and Regional Geosciences*, 31 (1), 57–65. doi:10.1016/S0016-7185(99)00044-5

Christoff, P., 1996. Ecological modernization, ecological modernities. *Environmental Politics*, 5 (3), 476–500. doi:10.1080/09644019608414283

Codd, J.A., 1988. The construction and deconstruction of educational policy documents. *Journal of Education Policy*, 3 (3), 235–247. doi:10.1080/0268093880030303

Convery, F.J., 2009. Origins and development of the EU ETS. *Environmental & Resource Economics*, 43 (3), 391–412. doi:10.1007/s10640-009-9275-7

Curran, G., 2009. Ecological modernisation and climate change in Australia. *Environmental Politics*, 18 (2), 201–217. doi:10.1080/09644010802682569

Dryzek, J., 2005. *The politics of the earth: environmental discourses*. 2nd. Oxford, New York: Oxford University Press.

Endl, A. and Berger, G., (2014) The 7th environmental action programme: reflections on sustainable development and environmental policy integration. *Quarterly Report for the European Sustainable Development Network (ESDN)*. Vienna: ESDN.

EU, 2006. Regulation (EC) No 1907/2006. *Official Journal of the European Communities*, 49, L 396.

European Commission, 1973. Environmental action programme. *Official Journal of the European Communities*, 16, C112.

European Commission, 1977. Environmental action programme. *Official Journal of the European Communities*, 20 C139.

European Commission, 1983. Environmental action programme. *Official Journal of the European Communities*, 26, C46.

European Commission, 1987. Environmental action programme. *Official Journal of the European Communities*, 30, C328.

European Commission, 1993. Towards sustainability: a European community programme of policy and action in relation to the environment and sustainable development. *Official Journal of the European Communities*, 36, C138.

European Commission, 2000. *Environment 2010: our future, our choice: the sixth environmental action programme*. Communication from the Commission. Brussels: European Commission.

European Commission, 2012. *Energy roadmap 2015*. Luxembourg: Publications Office of the European Union.

European Commission, 2014. *General union environment action programme to 2020: living well, within the limits of our planet*. Luxembourg: Publications Office of the European Union.

European Commission, 2015. EU ETS handbook. Available at: ec.europa.eu/clima/policies/ets_en [Accessed 30 June 2018]

European Commission, 2016. *The EU emissions trading system fact-sheet*. Luxembourg: Publications Office of the European Union.

Fischer, S., and Gedden, O., 2015. The changing role of international negotiations in EU climate policy. *The International Spectator*, 50 (1), 1–7. doi: 10.1080/03932729.2015.998440

Foucault, M., 1991. Politics and the study of discourse. *In*: G. Burchell, C. Gordon, and P. Miller, eds. *The Foucault effect: studies in governmentality*. Chicago: University of Chicago Press, 53–72.

Glasson, B., 2012. Gentrifying climate change: ecological modernisation and the cultural politics of definition. *MC Journal*, 15 (3). Available from: http://journal.media-culture.org.au/index.php/mcjournal/article/view/50

Hajer, M.A., 1997. *The politics of environmental discourse: ecological modernization and the policy process*. Oxford: Oxford University Press.

Hajer, M.A., 2005. Coalitions, practices, and meaning in environmental politics: from acid rain to BSE. *In*: D. Howard and J. Torfing, eds. *Discourse theory in European politics: identity, policy and governance*. Hampshire: Palgrave Macmillan, 297–315.

Hewitt, S., 2009. Discourse analysis and public policy research. *Centre for Rural Studies Discussion Paper Series*. (24). Available from: ippra.com/attachments/article/207/dp24Hewitt.pdf [Accessed 29 June 2018].

Hey, C., 2006. EU environmental policies: a short history of the policy strategies. *In*: S. Scheuer, ed. *EU environmental policy handbook: a critical analysis of EU environmental legislation*. Utrecht: International Books, 17–30.

Hillebrand, R., 2013. Climate protection, energy security, and Germany's policy of ecological modernization. *Environmental Politics*, 22 (4), 664–682. doi:10.1080/09644016.2013.806627

Hofmann, A., 2019. Left to interest groups? On the prospects for enforcing environmental law in the European Union. *Environmental Politics*, 28 (2).

Jaenicke, M., 2004. Industrial transformation between ecological modernisation and structural change. *In*: K. Jacob, M. Binder, and A. Wieczorek, eds. *Governance for Industrial Transformation. Proceedings of the 2003 Berlin Conference on the Human Dimensions of Global Environmental Change*. Berlin: Environmental Policy Research Centre, 201–207.

Jessup, B., 2010. Plural and hybrid environmental values: a discourse analysis of the wind energy conflict in Australia and the United Kingdom. *Environmental Politics*, 19 (1), 21–44. doi:10.1080/09644010903396069

Jordan, A., and Lenschow, A., 2000. 'Greening' the European Union: what can be learned from the 'leaders' of EU environmental policy. *European Environment*, 10 (3), 109–120. doi: 10.1002/1099-0976(200005/06)10:33.0.CO;2-Z

Judge, D., 1992. A green dimension for the European community? *Environmental Politics*, 1 (4), 1–9. doi:10.1080/09644019208414043

Laclau, E. and Mouffe, C., 2001. *Hegemony and socialist strategy: towards a radical democratic politics*. London: Verso.

Lekakis, J.N. and Kousis, M., 2013. Economic crisis, troika and the environment in Greece. *South European Society and Politics*, 18 (3), 305–331. doi:10.1080/13608746.2013.799731

Machin, A., 2013. *Negotiating climate change: radical democracy and the illusion of consensus*. London: Zed.

Maniates, M. and Meyer, J., 2010. *The environmental politics of sacrifice*. Cambridge, MA: MIT Press.

Meadows, D., et al., 1972. *The limits to growth*. London: Pan.

Meckling, J. and Jenner, S., 2016. Varieties of market- based policy: instrument choice in climate policy. *Environmental Politics*, 25 (5), 853–874. doi:10.1080/09644016.2016.1168062

Mol, A.P.J. and Sonnenfeld, D.A., 2000. Ecological modernisation around the world: an introduction. *Environmental Politics*, 9 (1), 3–16. doi:10.1080/09644010008414510

Mol, A.P.J. and Spaargaren, G., 2000. Ecological modernisation theory in debate: a review. *Environmental Politics*, 9 (1), 17–50. doi:10.1080/09644010008414511

Oberthuer, S. and Roche Kelly, C., 2008. EU leadership in international climate policy: achievements and challenges. *The International Spectator*, 43 (3), 35–50. doi:10.1080/03932720802280594

Pepper, D., 1998. Sustainable development and ecological modernisation: a radical homocentric perspective. *Sustainable Development*, 6, 1–7. doi:10.1002/(SICI)1099-1719(199803)6:1<1::AID-SD83>3.0.CO;2-8

Remling, E., 2018. Depoliticizing adaptation: a critical analysis of EU climate adaptation policy. *Environmental Politics*, 27 (3), 477–497. doi:10.1080/09644016.2018.1429207

Ruser, A., 2015. By the markets, of the markets, for the markets? Technocratic decision making and the hollowing out of democracy. *Global Policy*, 6 (1), 83–92. doi:10.1111/gpol.2015.6.issue-S1

Schuppert, F., 2011. Climate change mitigation and intergenerational justice. *Environmental Politics*, 20 (3), 303–321. doi:10.1080/09644016.2011.573351

Skovgaard, J., 2014. EU climate policy after the crisis. *Environmental Politics*, 23 (1), 1–17. doi:10.1080/09644016.2013.818304

Stavins, R., 2008. Cap-and-trade or a carbon tax. *The Environmental Forum*, 26 (5), 16.

Stephan, B. and Paterson, M., 2012. The politics of carbon markets: an introduction. *Environmental Politics*, 21 (4), 563–582. doi:10.1080/09644016.2012.688353

Szarka, J., 2012. Climate challenges, ecological modernization, and technological forcing: policy lessons from a comparative US-EU analysis. *Global Environmental Politics*, 12 (2), 87–109. doi:10.1162/GLEP_a_00110

Tienhaara, K., 2014. Varieties of green capitalism: economy and environment in the wake of the global financial crisis. *Environmental Politics*, 23 (2), 187–204. doi:10.1080/09644016.2013.821828

Toke, D., 2001. Ecological modernisation, social movements and renewable energy. *Environmental Politics*, 20 (1), 60–77. doi:10.1080/09644016.2011.538166

Ward, S., 1997. The IGC and current state of EU environmental policy: consolidation or roll-back? *Environmental Politics*, 6 (1), 178–184. doi:10.1080/09644019708414317

Warner, R., 2010. Ecological modernisation theory: towards a critical ecopolitics of change? *Environmental Politics*, 19 (4), 538–556. doi:10.1080/09644016.2010.489710

Weale, A., *et al.*, 2000. *Environmental governance in Europe: an ever closer union?* New York: Oxford University Press.

Weale, A. and Williams, A., 1992. Between economy or ecology? The single market and the integration of environmental policy. *Environmental Politics*, 1 (4), 45–64. doi:10.1080/09644019208414045

Zito, A., Burns, C., and Lenschow, A., 2019. Is the trajectory of European Union environmental policy less certain? *Environmental Politics*, 28 (2).

The importance of compatible beliefs for effective climate policy integration

Katharina Rietig

ABSTRACT
European climate policy faced increasing constraints during the economic and Eurozone crises (2008–2014). The European Commission subsequently refocused policymaking toward integrating climate objectives into other policy areas such as energy and the 2014–2020 European Union (EU) budget. The conditions for successful climate policy integration (CPI) are analyzed, focusing on the compatibility of key actors' beliefs. In renewable energy policy, CPI was successful as long as the co-benefits and related policy-core beliefs of energy security, rural economic development and climate action coexisted harmoniously. Once conflict among these policy-core beliefs emerged during the biofuels controversy, CPI was weakened as actors with competing economy-focused beliefs controlled the decision-making process. The case of EU budget climate mainstreaming illustrates how actors can add climate objectives into legislation despite meaningful discussion being 'crowded out' by other priorities. The findings highlight the importance of low conflict between departments, compatible beliefs and policy priorities for successful CPI.

Introduction

The European Union (EU) took an internationally leading position as a laboratory for developing progressive climate policies (Wurzel and Connelly 2011) between 2005 and 2010. The EU designed a number of climate policies such as the European Emission Trading Scheme (EU 2003) and pledged to reduce greenhouse gas emissions by 20% by 2020 (compared to 1990 levels), increase the share of renewable energies to 20% and improve energy efficiency by 20% (EC 2008). Yet, in the wake of the financial/ economic/ Eurozone crises 2008–2013, European leadership on climate change evolved into a more moderating role (Skovgaard 2014, Slominski 2016) as economic concerns began crowding-out political support for stronger climate policies. Key actors at the European Commission (EC), especially the Cabinet and Directorate General (DG) for Climate Action (CLIMA) (until 2010 part of DG Environment), saw declining opportunities to strengthen climate action via policies solely focused on addressing climate change. At the same time, there was increasing pressure from the United Nations Framework Convention on Climate Change (UNFCCC) negotiations to honor Europe's leadership role by legislating and implementing ambitious climate action (Wurzel and Connelly 2011, Lindenthal 2014). Civil servants at Cabinet/ DG CLIMA subsequently focused their attention on integrating climate change objectives into other policy areas such as energy, agriculture, transport or the EU budget. We can understand this 'climate mainstreaming' as climate policy integration (CPI) (Rietig 2013).

We have a limited understanding of when, why and under what conditions CPI can be effective, i.e. strengthen climate action, and when it is more likely to remain lip-service and unsuccessful in reducing emissions or improving societies' climate adaptation capabilities. Here, I argue that the successful integration of climate objectives into other policy areas (i.e. CPI) requires that CPI meets a number of conditions on policy coherence and coordination. However, our understanding of the conducive conditions for effective CPI remains limited, requiring a better understanding of the conditions for CPI to emerge as policy outcome. Especially in times of economic austerity, when the political agenda can potentially be preoccupied with other issues and pays little attention to climate policies, actors can facilitate climate action via CPI by focusing on areas requiring few additional financial resources and providing co-benefits for both achieving the sectoral policies' objectives and addressing climate change. Here, I analyze the conditions for CPI to emerge when political support for climate policies is slipping away. The central research questions are as follows: What role can CPI play in advancing climate action? and What conditions determine whether climate objectives are integrated into other policy areas?

Answering these questions furthers our understanding of the underlying factors driving policy change in European environmental policy, addressing wider questions concerning the degree to which environmental policy has significantly changed, the key external and/ or internal drivers for policy change and its implications for environmental policy and EU studies (Zito et al. Forthcoming – this volume). To answer the two research questions, I focus on two illustrative case studies that the Barroso Commission negotiated during the economic and Eurozone crises (2008–2013): the 2009 European Renewable Energy Directive (RED); and dedicating 20% of the €959.99 billion 2014–2020 EU budget (the Multiannual Financial Framework [MFF]) to policy measures co-benefitting climate action. These cases are particularly relevant as they are important elements of the EU's climate policy, especially the RED, which was a key component of the 20-20-20 Climate Package. The EC also reformed the biofuels aspects of the 2009 RED (EU 2009), which shares its sustainability criteria with the 2009 Fuel Quality Directive. The RED and MFF cases exemplify key CPI challenges: Cabinet/ DG CLIMA was not in control of the policy proposals but needed to negotiate with other Cabinets/ DGs, while both pursued their own objectives. The next section reviews the conditions for successful CPI. The empirical section examines the policymaking dynamics during the RED (2008/2009) and MFF (2011–2013) negotiations. Key actors at Cabinet/ DG CLIMA refocused their strategies toward 'climate mainstreaming' to continue to advance their climate action agenda despite reduced political attention and support. I subsequently discuss the conditions for CPI to occur, and finally present the key finding about compatible underlying beliefs.

Conditions for successful CPI

CPI is distinct from environmental policy integration (EPI). The literature defines EPI as a governing process or policy outcome that integrates environmental objectives into all other policy areas across policymaking stages, including planning, implementation and evaluation (Lenschow 2002, Lafferty and Hovden 2003, Nilsson and Eckerberg 2007, Jordan and Lenschow 2010). CPI is a governing process or policy outcome that integrates climate change objectives into non-climate change policy areas across all stages of the policy cycle, while CPI and EPI ideally overlap in 'sustainable CPI' (Rietig 2013), so that three EPI objectives are also applicable to CPI: the achievement of sustainable development and preventing damages to the environment, removing contradictions within and between policies and the realization of mutual benefits (Collier 1997, p. 36).

There is no consensus in the literature on whether CPI is an element of EPI or whether they are related but distinct concepts (Runhaar et al. 2014, Humalisto 2015, Dupont 2016). I conceptualize CPI as related but separate from EPI. CPI and EPI overlap in many regards and ideally sustainable CPI is as similar to EPI as possible (Rietig 2013). However, although climate change is an environmental challenge, it differs from most geographically limited environmental problems. CPI also has no status of principled priority, nor is it enshrined into EU treaties (Adelle and Russel 2013). Climate change is especially difficult to address effectively due to financial, technological, institutional, behavioral, governance and collective action-related challenges, requiring coordinated efforts across policy areas and governance levels (Jordan et al. 2010). Subsequently, objectives to address climate change via mitigation and/ or adaptation cannot be integrated, or 'mainstreamed' (Remling 2018), into all policy sectors with the same success. The central determinant for successful CPI is the level of synergies between the sector's policy objectives and climate objectives (Koch and Lindenthal 2011). Synergies can exist in terms of policy objectives, technology, innovation and infrastructure. They depend on the potential for climate change mitigation/ adaptation and how easily climate objectives and the other policy sector's objectives can be harmonized via regulatory instruments. Furthermore, there is no consensus about how to conceptualize climate change as an environmental problem. The complex interdependencies of policy areas require coordination and constructive collaboration to effectively address climate change and develop integrated policies (Cohen et al. 1998).

The 'wicked' nature of climate change and the cross-sectoral approach of CPI warrant a closer analysis of the conditions conducive to achieving policy change in the form of 'successful' CPI. I define CPI as successful if the policy outcome has co-benefits both for achieving the objective of the sectoral policy (e.g. energy provision, food security) and reducing greenhouse gas emissions (climate mitigation) and/ or improving climate change adaptation. The public policy literature identified policy entrepreneurs, the coherence of underlying beliefs among key actors and the co-benefits of policy objectives as central conditions for policy change (Baumgartner and Jones 1993, Kingdon 1995, Weible and Sabatier 2009). Policy entrepreneurs play an important role in affecting policy change and subsequently facilitating or hindering CPI. They are dedicated and highly skilled actors who define problems, are well connected with relevant networks, build coalitions to further their cause and take on central leadership roles to influence the political decision-making process (Braun 2009, Mintrom and Norman 2009, p. 651).

Actors hold different beliefs guiding their preferences and behavior, which authors differentiate into deep-core, policy-core and secondary beliefs (Weible and Sabatier 2009). Deep-core beliefs describe very stable fundamental worldviews, e.g. political left/ right orientations or green beliefs about the intrinsic value of nature. Policy-core beliefs result from the normative interpretation of a policy problem taking into account an actors' deep-core beliefs, e.g. whether one should address climate change using market-based policy instruments. They tend to be relatively stable and also unlikely to change quickly. Secondary beliefs describe actors' preferences for specific policy instruments (e.g. emission trading versus taxes among market-based instruments and actions in line with their fundamental worldviews/ deep-core beliefs). They change over time as additional scientific evidence becomes available, societal preferences evolve or political framework conditions shift. Different actors advocate certain 'solutions' to the problem addressed in the policymaking process (Weible and Sabatier 2009).

While there may be countless different perspectives, actors tend to cluster together in groups and collaborate in a coordinated way with other actors sharing similar normative and causal beliefs to leverage power in the democratic political process dominated by the need for (voting) majorities. Membership in such coalitions advocating certain policy solutions is open to all actors holding similar beliefs including interest groups, government representatives, legislators and members of expert networks (Mintrom and Norman 2009). They use different strategies to win over groups with competing interests and achieve decisions by governmental authorities in line with their beliefs. In the case of conflict, it is useful to understand interest groups as coalitions, whereby the policy outcome ultimately depends on which of these coalitions hold more political power (Nohrstedt 2011). Policy change can result from varying actor constellations, depending on how beliefs are aligned and compatible. Interest groups either collaborate as one group's policy-core/ secondary beliefs shift to match with the other groups' policy-core/secondary beliefs (Weible and Sabatier 2009, Lodge and Matus 2014). This creates co-benefits between two previously unrelated policy areas. It is also possible that groups of actors who previously cooperated due to seemingly aligned policy-core/ secondary beliefs on policy instruments can begin to oppose each other due to changes in policy-core or secondary beliefs (e.g. resulting from new scientific evidence removing previously perceived co-benefits between policies), while the deep-core beliefs remain unchanged (Rietig 2018). This is particularly relevant for policy integration where policy objectives are integrated into another groups' policy domain. Depending on how well the integrated objectives fit with the beliefs of the 'receiving' policy area, policy integration can be either successful, resulting in strengthened policy outputs, or remain weak, involving lip-service without achieving policy change that matches the integrated policy's objectives (Koch and Lindenthal 2011; Rietig 2018). Wider framework conditions also influence the success of CPI. These include societal demands, sudden external shocks, public opinion (Baumgartner and Jones 1993, Kingdon 1995) and how policymakers perceive and interpret voter demands with regard to current policy initiatives. Examples include rising public concern on climate change and the March 2011 Fukushima nuclear disaster that resulted in Germany's abandonment of nuclear energy (Jahn and Korolczuk 2012).

This discussion suggests two central conditions for CPI to be successful: underlying beliefs of policymakers matter for the level of conflict; and secondary and policy-core beliefs among key policymakers and their organizations must align for co-benefits to be identified and realized. The following sections trace these conditions in the empirical case studies on integrating climate objectives into European energy policy and the EU budget during the Barroso Commission and financial/ economic/ Eurozone crises (2008–2013). The energy case study focuses on developing the European RED during the Barroso Commission at the peak of the economic and Eurozone crisis as European flagship legislation on CPI. The EU budget case study examines the 'climate mainstreaming' approach of dedicating 20% of the 2014–2020 MFF to measures with climate action co-benefits. The MFF case is highly relevant since we can view it as a policy innovation that, if successfully transferred to national budgets, could significantly strengthen climate action also during times of financial or economic crises. The RED is an important part of Europe's flagship policies to address climate change, but developed out of 'classic' energy policy. The impetus came from the 1970s policy objective to improve Europe's energy security with 'alternative' energies and to support rural economic development with increased use of biofuels. The biofuels component also closely links to the EU's Common Agricultural Policy and subsidies (Sharman and Holmes 2010). The RED is particularly relevant to better understand the conditions for successful CPI in the EU: its three core objectives of energy security, economic development and climate action provide an illustrative case of the opportunities and challenges faced when integrating climate objectives into other sectoral policies, especially once the objectives no longer co-benefit each other but start to compete, and key actors prioritize them differently.

Methodology

The data set underpinning the empirical analysis includes 43 elite interviews with key actors involved in both case studies, and numerous policy documents including legislative proposals, directives, white/ green papers and communications from the European Parliament (EP), Council and Commission. The interview questions encouraged actors to identify their beliefs (deep-core/ policy-core and secondary beliefs), clarifying whether beliefs changed and why. Following transcription, I coded the interviews along the belief typology and analyzed the data with a process-tracing approach that followed the negotiation process of the RED and MFF (Hall 2013) using the qualitative analysis software program NVivo. The key respondents work(ed) in the EC, European member state (MS) governments and representations, the EP and stakeholder organizations such as environmental NGOs (ENGOs) and industry. The interviews include the civil servants working at the Directorates General and Cabinets for Energy (until 2009 DG Transport and Energy [DG TREN]), Environment [split in 2009 into DG Environment and DG CLIMA] and Agriculture and Rural Development [DG AGRI]. All four DGs (Energy, Environment, CLIMA and Agriculture) were involved in the EU MFF negotiations and multiple aspects of EU renewable energy via the RED (led by DG Energy/ TREN). The biofuel aspects illustrate the complicated nature of CPI in practice. This case study allows us to trace in-depth conditions for policy integration. While it is not possible to generalize from a limited number of case studies, the MFF negotiations and the RED development nevertheless offer an insight into the conditions that matter for CPI to emerge as a policy outcome (and thus as policy change).

Integrating climate objectives into energy policy during the Barroso commission

Negotiating the RED: aligned policy-core beliefs

Motivated by the oil shocks of the 1970s, the EU increased research and development into 'alternative energies' to improve energy security. The energy security benefit of renewables was reframed in the 1990s as the double objective of energy security and economic development with the added benefit of improving European integration via advancing the integrated electricity market. Following the rise of climate change on the international agenda, especially via the 1992 Rio Summit and the 1997 Kyoto Protocol, climate change emerged as a third objective. The means for addressing this objective was via an increased share of renewable energies (see Hildingsson et al. 2012). This triangle of energy security, economic development and climate action proved a 'magic formula' aligning with MS' different domestic interests, as well as cities and other stakeholder's objectives (Kalafatis 2018). Actors' policy-core beliefs across different interest groups were aligned (interview EC2), as priorities guiding their focus in political decision-making matched their 'material' interests for these three priorities. The RED adoption in the areas of electricity and heating/ cooling was relatively straightforward as policy-core beliefs were aligned in the 'magic formula' of energy security, economic development and climate change (Rietig 2018). Actors focused on deep-core beliefs regarding the priority of economic development held the policy-core belief that the RED should improve rural economic development and energy security. Simultaneously, actors with climate change and environmental priorities (and corresponding deep-core beliefs) saw the RED as central climate policy instrument within the EU Climate Package (Rietig 2018).

There was an alignment of policy-core beliefs in the form of a general consensus among European-level decision-makers on the desirability of increasing the share of renewables. This prompted the development of the 2001 Renewable Electricity Directive, which required MS to set national indicative targets (EU 2001, Article 3). The overall consensus (via aligned policy-core beliefs among stakeholders and decision-makers) was that alternatives to fossil fuels should play a larger role in the European energy mix, facilitating the Directive's adoption (EC1, EC3, MS3). The March 2006 Council meeting set the overall rationale for developing a renewable energy strategy (Council of the EU [Council] 2006), pp. 13–15 that matched the three policy-core beliefs – the need to improve energy security, address climate change and accelerate the uptake of renewables for an integrated energy market benefitting economic development. The EP passed a resolution requesting the EC to submit a legislative proposal for concrete measures on renewable energies for heating and cooling by 31 July 2006 (European Parliament 2006). Shortly thereafter the EC received a request to prepare a proposal that integrated all three aspects of renewables, resulting in the Renewable Energy Road Map (EC 2007). The Road Map already contained the landmark targets on renewable energy adopted at the March 2007 Council meeting (Council 2007, EC2) in the form of a legally binding 20% renewable energies target including a mandatory minimum 10% target for biofuels in 2010 (Rietig 2018). Achieving these targets required substantial strengthening of the EU regulatory framework (EC 2007, p. 18), i.e. a new directive. The January 2007 Renewable Energy Road Map had significant impact on the Council decisions and the Commission's subsequent 2008 RED proposal, which passed unusually quickly (EC2, EC3). Part of the large Climate Package, its central motivation was the need to deliver on the EU's Kyoto Protocol commitments and to demonstrate leadership in the upcoming climate change summit in Copenhagen 2009 (EC1, EC3). This allowed an unusually quick decision before the window of opportunity, facilitated by economic prosperity and EU enlargement, closed with the 2008/2009 economic crisis:

> We had an incredible political momentum (...) [The RED] enjoyed support from the Council, from Barroso personally and so it was part of a bigger vehicle that was very hard to stop. (...) It [was] pushed through as one of their prizes of the French Presidency, so it was agreed actually some months after the big crisis. (EC3)

Nevertheless, among Cabinet/ DG Environment/ CLIMA and Cabinet/ DG Energy/ TREN, key secondary beliefs were conflicting especially regarding how to achieve the target and what commitments it would require from MS. The political debate focused on whether targets should be indicative or mandatory, and on MS' share of renewable energies (MS1). The RED also experienced an unusually quick adoption due to the Commission's successful pre-negotiation process prior to publishing the official proposal containing a formula acceptable to the MS (EC6, EC9, Member of European Parliament [MEP] 2). Interview data across actor groups suggest that the EC and, in particular, DG TREN, which was responsible for the proposal, were key drivers and policy entrepreneurs behind the renewable energy targets. For example, in the 1990s/2000s the Commission established and promoted networks supporting the uptake of renewable energies on the local level, such as FEDEREN and the Covenant of Mayors (EC1). These local initiatives facilitated mutual learning and enabled local authorities to play a more active role in climate action (Domorenok, Forthcoming).

In conclusion, several factors explain why the 2009 RED was adopted unusually quickly and how it became a central policy for successfully integrating climate change objectives into energy policy. The policy-core beliefs of actors were aligned as long as they were able to focus on the co-benefits of renewable energies for addressing climate change and thus delivering on international commitments (excluding the biofuels controversy discussed in the next section), improving energy security and strengthening economic development. It was the result of a four-decade long EU-internal development process that started with the 1970s oil shocks and worries about energy security, and continued with the adoption of two previous EU directives on renewable electricity and biofuels. It was also part of the larger 20-20-20 Climate Package (EC 2008), which increased the political pressure to reach agreement.

Ineffective CPI when co-benefits are challenged and competing beliefs emerge

The RED legislative proposal also contained three aspects that became the subject of major disagreement among actors involved. While the question regarding individual MS targets was resolved before the proposal was published (EC2, EC6), the issue regarding trade in green electricity certificates (guarantees of origin) was subject to major disagreement between EC DGs, MS, ENGOs and industry lobbying groups (Toke 2008) due to their negative implications for stable investment decisions. While these disagreements concerned distributive issues among MS, the controversy regarding the mandatory 10% target on renewable energies in transport (EC9, EC11, EU 2009) was more fundamental.

The role of beliefs is particularly central to understanding how the biofuels controversy resulted in weakened CPI. Policy-core beliefs regarding the importance of economic development and energy security guided central actors at DG Energy, in the MS and the biofuels industry when they designed renewable energy policy (EC1, EC2, EC8, EC9, MS2, Industry 2). This also reflects the historic path-dependence generated by the 1970s alternative energy objective and using biofuels for transport to improve energy security and reduce energy dependence from non-EU countries (EC1, MS3, Industry 3). DG Environment/ CLIMA, ENGOs and some actors in the EP and MS held deep-core beliefs in the principled priority of environmental protection and climate action (EC5, EC9, EC10, EC11, MEP4, MEP5). As long as all actors shared the policy-core belief that all renewable energies, including biofuels, were beneficial for addressing climate change, both the environment-focused and the economic development-focused coalitions (Rietig 2018) were pursuing the same policy objectives. They were united in their policy-core belief about increasing the share of renewable energies, including biofuels. As the DG responsible for the RED, representatives of DG Energy/ TREN played a central role in getting the European Council to agree to the renewables targets (EC2, EC8, EC9, Sharman and Holmes 2010). They used policy entrepreneurial strategies, such as taking on strong leadership roles advocating the importance of the policy proposal for delivering on existing policy commitments; they ultimately convinced the Council, to request a Commission proposal containing a 10% biofuel target (EC3, Sharman and Holmes 2010). DG Energy/ TREN thus pre-committed to this target, making it difficult later to change course, lobbying central actors in MS for their support and selectively using and commissioning supporting scientific evidence (Palmer 2015). Once competing adverse scientific evidence on the negative climate impacts of food-crop-based biofuels and their negative indirect land use change effects emerged (Searchinger *et al.* 2008), both coalitions refocused on their deep-core beliefs in support of energy security and economic development in the case of DG Energy, and on the priority of climate action in the case of DG Environment/ CLIMA (Rietig 2018). As DG Energy remained responsible for the RED, the biofuels target effectively remained in the form of 10% of renewable energies in transport given the limited alternatives to food-crop-based biofuels (Sharman and Holmes 2010, Palmer 2015). DG Environment/ CLIMA succeeded in integrating a limited number of environmental safeguards into the RED via the Fuel Quality Directive in the form of the sustainability criteria for biofuels and the requirement for a review of the biofuels content, especially with regard to indirect land use change effects (EC9, EC10, ENGO1, ENGO2).

The biofuels controversy explains why CPI can result in major controversies and, depending on the power that the involved coalitions exercise, even in 'green-washing' or adverse effects for climate action and environmental protection. CPI is full of institutional pitfalls once disagreements emerge between actors based on incompatible beliefs and political objectives. The central reason for the biofuels controversy over the positive or negative climate impacts of food-crop-based biofuels was the contestation over scientific knowledge at the time of policymaking. Policy lock-ins exacerbated this controversy once the Council agreed to the target (Council of the EU [Council] 2006), allowing for only incremental steps to 'correct' the policy outcomes that were made under scientific uncertainty. Furthermore, in the meantime MS invested in an industry focusing on food-crop-based biofuels. Later reforms of biofuels policy limiting further increases in the share of food-crop-based biofuels came with significant economic costs and loss of trust in policymaking, especially for the 'Central and Eastern European countries, they haven't had the same boom in (...) wind, but biofuels they've done really well in so I think it's quite frustrating then to have the rug sort of swept out from under their feet' (MS4). The policy proposal on indirect land use changes put forward in 2012 and legislated in 2015 mitigated the worst consequences, but did not take a strong precautionary approach to biofuels as environmentally focused actors inside and outside the EC, in particular DG Environment/ CLIMA, demanded. It rather continued with incremental changes to the business-as-usual status quo. In particular, the EC proposed limiting the amount of food-crop-based biofuels and bioliquids that can be counted toward the 10% target to the existing consumption level of 5% (EC 2012, Article 2(2c)ii).

This effectively meant that the remaining 5% of renewable energies in transport would have to come from second-generation (non-food-crop-based) biofuels or they would not count toward the overall target. It also included incentives for electric cars and especially second/ third generation biofuels with no or low indirect land use change effects. These included a focus on longer types of straw, different types of waste and algae (EC 2012, EC4, Industry1). This outcome can be regarded as resembling CPI, although CPI could have been strengthened by reducing the existing share of food-crop-based biofuels.

The policy-core beliefs of Cabinet/ DG Energy/ TREN and Cabinet/ DG Environment/ CLIMA conflicted as it became clear that the policy-core beliefs of Cabinet/ DG Energy/ TREN prioritized energy security and rural economic development, while Cabinet/ DG Environment/ CLIMA's policy-core beliefs prioritized environmental protection and climate action. This conflict between the Cabinets/ DGs subsequently hindered successful CPI. The shared competencies between Cabinet/ DG Environment/ CLIMA, which was responsible for the Fuel Quality Directive, and Cabinet/ DG Energy, with a lead on the RED, meant that, once cooperation was difficult, it came down to Cabinet/ DG Energy to make use of the existing leadership position and 'ownership' of the policy proposal (EC3), including the use of policy entrepreneurial strategies to get its policy proposal adopted (Sharman and Holmes 2010, Palmer 2015). Energy Commissioner Piebalgs (2005–2009) strongly supported mandatory biofuels targets in the RED (EC6). The 2010–2014 Energy Commissioner Günther Öttinger further weakened CPI; he advocated an indicative renewable energy target for 2030 instead of using the opportunity to push for a stronger mandatory target (Bürgin 2015):

> There is some discontinuity, that is the new Commissioner has different ideas from the previous Commissioner. What I can see [is that the] policy approach is changing with the person. (…) What we have now is (…) [the] Commissioner [for Climate Action] openly advocating a renewable energies target for 2030 and [Energy] Commissioner Öttinger not going as far. (EC6)

The institutional reorganization, following the change in leadership of the EC, further weakened CPI after 2014. The 2014 Juncker Commission reorganized both the Environment and the Climate Action portfolios, integrating Climate Action with Energy, and Environment with Maritime Affairs and Fisheries. In both cases, the policy-core beliefs and corresponding organizational priorities of economic development and energy security stood at odds with organizational priorities of environmental protection and climate mitigation. Where actors cannot identify or where they contest co-benefits, environmental protection and climate mitigation are likely to take a back-seat to the declared priority of economic recovery and addressing external security challenges (Čavoški 2015).

Mainstreaming climate action into the EU budget during the Eurozone crisis

In response to reduced political support for stronger climate policies during the financial/ economic/ Eurozone crises (Bürgin 2015, Slominski 2016), Cabinet/ DG CLIMA focused on mainstreaming climate action into other sectoral policy areas. In 2011, the EC introduced the proposal to dedicate 20% of the EU 2014–2020 budget to CPI measures (EC 2011), as it saw this form of CPI as the best option to mirror the 20-20-20 Climate Package in the EU budget (EC15). This corresponded with Cabinet/ DG CLIMA's policy-core beliefs in strengthening climate action and making use of opportunities across the EC to achieve the corresponding climate action objective.

Coordination and cooperation between Cabinet/ DG CLIMA and other Cabinets/ DGs was crucial (Koch and Lindenthal 2011) as CPI was not a policy proposal in its own right, but consisted of interventions into other DGs' policy domains. When CPI conflicted with sectoral policy-core beliefs on what policies to prioritize, the attempt to integrate climate objectives frequently resulted in resistance on the policy-drafting levels (EC14). These were not resolved between DGs within the Commission due to incompatible policy-core beliefs about the relevance and importance of CPI (EC10, EC11, EC15), but also due to the lack of agreement in consultations between DGs carried 'up the hierarchy' into the College of Commissioners meeting. Cabinet/ DG CLIMA ultimately made use of the political opportunity and the Commissioner for Climate Action succeeded at persuading the other Commissioners, so that 'it was literally in the College meeting where it was decided' (EC15) to include the 20% climate mainstreaming objective. This illustrates how CPI entered a policy proposal due to policy entrepreneurial activities of actors holding climate action-focused deep-core and policy-core beliefs. Cabinet/ DG CLIMA did not however change the beliefs of other actors; other actors across the Commission did not engage in learning that CPI is important, but used conventional bargaining and negotiation tactics (Rietig and Perkins 2018).

Consequently, the EC's MFF proposal contained the 20% climate-mainstreaming objective, which however remained a policy-core belief only a limited number of actors within the EC shared. Other than the biofuels target in the RED, the 20% mainstreaming objective again attracted little attention during the negotiations between the Commission, Parliament and Council. It rather 'slipped through' with little discussion as MS, MEPs and other stakeholders were more concerned with economic development issues and the funds dedicated to agriculture, transport and cohesion. As several actors involved in the negotiations emphasized:

> In all the discussions, the EU, I don't think I ever really heard anybody in the CAP [Common Agricultural Policy] reform negotiations refer to the 20%

[climate mainstreaming]. I mean I think it's clear that in the process member states were very good at saying 'yes we all need to deliver real environmental benefit', but then we spent a lot of time actually trying to limit the impact of the greening proposals. (MS10)

DG CLIMA understands the 20% climate mainstreaming objective as a set of indicators helping to determine the extent to which a sectoral policy such as the CAP also meets climate change objectives in addition to food security as key objective:

> The real revolution of the CAP reform might be totally invisible. It's the [climate mainstreaming] indicators. Measuring the CAP's success by indicators means we have to look for certain results and we have to put them in figures (…). Maybe these indicators will play a much bigger role. (EC19)

Successful CPI would have required reflection by MS and DGs about how to implement or strengthen this approach in national budgets. Yet, there was no discussion of mainstreaming in the MFF negotiations. Such discussion would have required MS to form a position on the issue by reflecting on their national interests and determining to what extent the climate mainstreaming objectives matched their deep-core, policy-core and secondary beliefs. It would have raised the awareness of policymakers who usually do not deal with climate change and also presented an opportunity to learn (Rietig and Perkins 2018). However, by occupying a more prominent place in the budget negotiations, the climate mainstreaming objective might have suffered deletion or having its percentage target reduced; this was a major concern among key actors at Cabinet/ DG CLIMA in the run up to the MFF EU budget negotiations in the Parliament and Council after their difficult experiences with other Cabinet/ DGs during the EC internal MFF proposal negotiation process.

> I do really feel there is a conservatism [against increasing climate action] in the mentality of the Ministers of Finance and the EU budget is primarily determined by them, but equally the conservatism exists in the Ministries of Agriculture and Cohesion, you know we've got to change thinking and that is a difficult job, but we keep working at it. (…) We did extremely well. But it was the high point. I think 2011 was an extremely good year for mainstreaming in the Commission. But I also used a lot of political capital getting it. And I am now the most unpopular guy in Brussels. (…) Because I am interfering with other people's portfolios, telling them how to do their job. People don't like that. So it's difficult. (EC15)

This reflection from a high-ranking policymaker at the center of the negotiations on climate mainstreaming in the EU budget points to the core challenge and dilemma of successful CPI: if secondary or especially policy-core beliefs of policymakers in the 'receiving' policy area are less sympathetic toward climate action, then achieving climate objectives rests upon political power, bargaining tactics, policy entrepreneurial strategies and political 'horse-trading'. If, however, economic/ financial or other crises elevate other policy objectives such as economic development, food and energy security above climate change concerns, it is difficult for CPI to be successful. CPI either gets 'watered down', as was the case in the RED and biofuels controversy, or receives little attention in the negotiation phase as the 20% climate mainstreaming in the EU budget case illustrated. In this case, the underlying beliefs of sectoral policymakers and climate-focused policymakers did not clash openly as the sectoral policymakers hardly noticed that CPI had occurred. Instead, it 'slipped' into the EC's MFF proposal due to policy entrepreneurial tactics in the College of Commissioners meeting; the negotiations between the Parliament and Council did not remove it as policymakers' attention was focused on more immediate and 'important' economic and financial aspects. Subsequently, CPI made it into the policy outcome (i.e. the MFF), but the implementation of climate action became more difficult when it rested upon integration into sectoral policies that went beyond easily achievable co-benefits. This opened new areas of potential conflict once incompatible policy-core beliefs emerged (Koch and Lindenthal 2011).

Discussion and conclusion on the conditions for CPI

I have discussed and illustrated the underlying factors for changes in EU environmental policy in the area of CPI. There has been a change in EU environmental policy, but it has been incremental; EU-internal factors relating to the extent to how well underlying beliefs among central actors were aligned (i.e. how actors view and assess the world) have driven this process (Zito *et al.* Forthcoming). Which actors dominate in turn determines the political decision-making arena and how it changes over time.

The empirical sections traced the conditions for successful CPI at the example of the RED and the EU budget. Both case studies showed that for CPI to be successful, underlying beliefs of policymakers mattered for the level of conflict, and that secondary/ policy-core beliefs among key policymakers and their organizations needed to be aligned for co-benefits to be identified and recognized. The findings have implications for our wider understanding of EU environmental policy and the ease of using CPI to address climate change (Zito *et al.* Forthcoming). The RED case examined how climate objectives were integrated into energy policies by emphasizing co-benefits in terms of energy security, rural economic development and addressing climate change. The co-benefits also matched the deep-core and policy-core beliefs of the policymakers involved that prioritized either energy security and economic development (Cabinet/ DG Energy/ TREN) or climate action (Cabinet/ DG Environment/ CLIMA). The RED catered to the corresponding policy-core and secondary beliefs (i.e. maximizing the amount of renewable energies). It is important to note that factors other than climate mitigation initially motivated renewable energy policy in the 1970s. Only in the 1990s and early 2000s, as climate change became a strong global concern and the EU needed to implement the Kyoto Protocol that entered into force in 2005, actors both inside and outside the EC reframed renewable energy as a contribution to climate mitigation via low emission energy production and increasing green vegetation as a carbon sink. In combination with an internal desire to advance renewable energies for energy security and rural development reasons, the EC used policy entrepreneurial activities to promote renewable energies. At the same time, the EC, MEPs, MS and non-state actors regarded the economic situation as favorable enough to allow 'low politics' such as climate change to enter the political agenda. The Fourth Assessment Report of the Intergovernmental Panel on Climate Change (IPCC 2007) provided the scientific and economic evidence to act on climate change earlier rather than later, and the upcoming UNFCCC negotiations in December 2009 in Copenhagen added external political pressure. These framework conditions facilitated CPI, but only as long as the actual co-benefits between climate action, energy security and economic development were present and the underlying deep-core/policy-core beliefs of policymakers were not conflicting (as the RED catered to all three benefits). Once new scientific knowledge, competing institutional competencies and objectives challenged these beliefs, CPI was weakened as incompatible underlying deep-core and policy-core beliefs emerged. This points toward the importance of compatible underlying deep-core/ policy-core beliefs between representatives of the climate and sectoral policy area. As long as secondary/ policy-core beliefs matched, there was a consensus in favor of the RED. Once the beliefs did not match, as climate action-focused actors changed their beliefs on the climate benefits of biofuels, secondary/ policy-core beliefs between actors and Cabinets/ DGs were no longer aligned, and CPI co-benefits could not be identified and recognized. Even more, the subsequent conflict between key policymakers resulted in a weakening of CPI in negotiations for the 2030 targets (Bürgin 2015). This matches the picture of varying ambition of actors involved in EU environmental policy and their respective constituencies ranging from policy entrepreneurial activities to decreased ambition and potential dismantling throughout the policy cycle (Hoffman Forthcoming – this volume, Schoenefeld and Jordan (Forthcoming) – this volume, (Steinebach and Knill 2017, Wurzel et al. Forthcoming) – this volume).

The second case study on mainstreaming climate action into the MFF by earmarking 20% of the MFF expenditures for climate objectives adds an additional element to the two conditions for successful CPI. Within the EC, the focus on economic issues and conflicting policy-core beliefs crowded CPI out of the MFF proposal until it was included at the College of Commissioner level due to policy entrepreneurial acumen. In the Parliament and Council negotiations, more 'important' issues again crowded CPI out, but it 'slipped through' into the final MFF policy outcome due to a lack of political attention and debate. Although climate mainstreaming emerged as a policy outcome, there was a lack of compatible beliefs about the importance of climate action. This can be understood as a missed opportunity for policy learning (Rietig and Perkins 2018) as more prominent discussions would have given MS the option to reflect on climate mainstreaming and its co-benefits – which subsequently could have resulted in policy transfer to national budgets. The second case study thus illustrates that CPI can remain weak when there is a lack of attention and discussion in the negotiations – actors therefore need to reflect on a policy proposal first to determine its compatibility with their policy-core beliefs. If there is a high compatibility, the level of conflict remains low and CPI can become a policy outcome with good prospects of being implemented (i.e. successful CPI).

These findings resonate with the drivers for policymaking emphasized by empirical studies concerning the importance of compatible deep-core and policy-core beliefs, e.g. between DGs within the EC in the area of EPI (Koch and Lindenthal 2011) and transport (Palmer 2015). Many interviewees stressed that the opportunity to strengthen climate policies such as the European Emission Trading Scheme (Skjærseth and Wettestad 2010) waned with the economic and Eurozone crises as MS were preoccupied with more immediate economic concerns including increasing unemployment, resulting in a strengthening of the economic development-minded coalition (see, on the UK, Carter and Jacobs 2014).

This resulted in a lower priority for climate change concerns as they were less tangible for many policymakers and voters confronted with threats to the survival of (environmentally polluting) industries and higher costs associated with integrating climate objectives. Deeper into the economic and Eurozone crises, the overall priorities of the ECs' leadership changed when new Energy Commissioner Öttinger (Bürgin 2015) and the 2014 Juncker Commission more strongly emphasized economic recovery (Čavoški 2015). Climate policy enjoyed high attention while there were few pressing economic or security problems. Sectoral policymakers were willing to integrate climate objectives into their policy areas as long as their primary objectives (i.e. policy-core beliefs) were not contested. However, political support, and especially attention, diminished with the economic/ Eurozone crises and emerging security/ migration crises between 2013 and 2015. In addition, there was a stronger shift in MS to right-wing parties with a focus on economy and security at the expense of climate/

environmental policy (e.g. Poland, Hungary) (Wurzel *et al.* Forthcoming). Given the EC's objective to publish policy proposals with a realistic chance of being accepted with minor changes, the diminishing appetite for seemingly costly climate-related policies, particularly in Eastern and Southern European MS, was a major hindrance (EC3, EC9, MS1, MS3, MEP5). Furthermore, the Juncker Commission offered little endorsement of the importance of environmental and climate policy at the expense of urgent 'high politics' security challenges in the geographic neighborhood of the EU (Čavoški 2015).

However, both cases also illustrate possibilities for 'saving' climate action throughout difficult political times when problems, which political decision-makers perceive to be more urgent, threaten to crowd out climate policies. Integrating climate objectives using policy windows when policy-core and secondary beliefs are aligned or, in their absence, resorting to policy entrepreneurial strategies in the decision-making process can ultimately facilitate and strengthen policy stability until political framework conditions are more favorable and windows of opportunity allow for further increasing climate action ambitions. By attracting little political attention throughout the MFF negotiations, the climate-mainstreaming element remained in the MFF proposal and was adopted. This provides climate mainstreaming with the political legitimacy required for implementation (because it is part of the EU budget). It also allowed for innovative approaches such as developing climate mainstreaming indicators to account for co-benefits between sectoral policies and climate action. Once such indicators make financial and policy-related co-benefits between climate action and sectoral policies tangible, they can contribute to changing sectoral policymakers' policy-core beliefs toward a higher awareness of climate action and willingness to tolerate CPI in their policy areas. This can ultimately result in changing beliefs toward a higher level of compatibility and thus more successful CPI.

Disclosure statement

No potential conflict of interest was reported by the author.

References

Adelle, C. and Russel, D., 2013. Climate policy integration: a case of déjà vu? *Environmental Policy and Governance*, 23 (1), 1–12. doi:10.1002/eet.v23.1

Baumgartner, F. and Jones, B., 1993. *Agendas and instability in American politics.* Chicago: University of Chicago Press.

Braun, M., 2009. The evolution of emissions trading in the European Union – the role of policy networks, knowledge and policy entrepreneurs. *Accounting, Organizations and Society*, 34 (3–4), 469–487. doi:10.1016/j.aos.2008.06.002

Bürgin, A., 2015. National binding renewable energy targets for 2020, but not for 2030 anymore: why the European Commission developed from a supporter to a brakeman. *Journal of European Public Policy*, 22 (5), 690–707. doi:10.1080/13501763.2014.984747

Carter, N. and Jacobs, M., 2014. Explaining radical policy change: the case of climate change and energy policy under the British Labour Government 2006–2010. *Public Administration*, 92 (1), 125–141. doi:10.1111/padm.12046

Čavoški, A., 2015. A post-austerity European Commission: no role for environmental policy? *Environmental Politics*, 24 (3), 501–505. doi:10.1080/09644016.2015.1008216

Cohen, S., Demeritt, D., and Robinson, J., 1998. Climate change and sustainable development: towards dialogue. *Global Environmental Change*, 8 (4), 341–371. doi:10.1016/S0959-3780(98)00017-X

Collier, U., 1997. *Energy and environment in the European Union*. Aldershot: Ashgate.

Council, 2007 March. Council conclusions requesting proposals for 20-20-20climate strategy. 7224/ 1/07.REV1/CONCL1. 8/9.3.2007. *Journal of the European Communities*, Brussels.

Council of the EU (Council), 2006. Presidency conclusions. 7775/ 1/06REV1/CONCL1. 23/24.3.2006. *Journal of the European Communities*, Brussels.

Domorenok, E., Forthcoming. Voluntary instruments for ambitious objectives? The experience of the EU covenant of Mayors. *Environmental Politics*, 28 (2). [This issue].

Dupont, C., 2016. *Climate policy integration into EU energy policy. Progress and prospects*. London and New York: Routledge.

EC, 2008. *20-20-20 by 2020. Europe's Climate Change Opportunity*. COM(2008) 30 final. 28.1.2008. Brussels.

EC, 2011. *A budget for Europe 2020*. COM(2011) 500 final. 29.6.2011. Brussels.

EC, 2012. *Proposal for a directive of the European Parliament and of the European Council amending directive 98/70/EC relating to the quality of petrol and diesel fuels and amending directive 2009/28/EC on the promotion of the use of energy from renewable sources*, COM/2012/595 final, 2012/0288 (COD)C7-0337/12. 17.10.2012. Brussels.

EC, 2007. *Renewable energy road map, renewable energies in the 21st century: building a more sustainable future*, COM(2006) 848 final. 10.01.2007. Brussels.

EU, 2001 October 27. Directive 2001/77/EC on the promotion of electricity produced from renewable energy sources in the internal electricity market. L283. *Journal of the European Communities*, Brussels.

EU, 2003 October 25. Directive 2003/87/EC establishing a scheme for greenhouse gas emission allowance trading within the Community and amending Council Directive 96/61/EC. L275/32. *Journal of the European Communities*, Brussels.

EU, 2009 June 5. Directive 2009/28/EC on the promotion of the use of energy from renewable sources. L140/16. *Journal of the European Communities*, Brussels.

European Parliament, 2006. European Parliament Resolution with recommendations to the Commission on heating and cooling from renewable sources of energy. 2005/2122(INI). 14.02.2006. *Journal of the European Communities*, Brussels.

Hall, P., 2013. Tracing the progress of process tracing. *European Political Science*, 12 (1), 20–30. doi:10.1057/eps.2012.6

Hildingsson, R., Stripple, J., and Jordan, A., 2012. Governing renewable energy in the EU: confronting a governance dilemma. *European Political Science*, 11 (1), 18–30. doi:10.1057/eps.2011.8

Hoffman, A., Forthcoming. Left to interest groups? On the prospects for enforcing environmental law in the European Union. *Environmental Politics*, 28 (2). [This issue].

Humalisto, N., 2015. Knowledge in Climate Policy Integration: how non-governmental organizations re-frame the sciences of indirect land-use changes for policy-makers. *Environmental Policy and Governance*, 25 (6), 412–423. doi:10.1002/eet.1692

IPCC, 2007. *Climate change 2007: synthesis report*. Geneva: IPCC.

Jahn, D. and Korolczuk, S., 2012. German exceptionalism: the end of nuclear energy in Germany! *Environmental Politics*, 21 (1), 159–164. doi:10.1080/09644016.2011.643374

Jordan, A., et al., 2010. *Climate change policy in the European Union. Confronting the dilemmas of mitigation and adaptation?* Cambridge: Cambridge University Press.

Jordan, A. and Lenschow, A., 2010. Environmental policy integration: a state of the art review. *Environmental Policy and Governance*, 20 (3), 147–158. doi:10.1002/eet.v20:3

Kalafatis, S.E., 2018. When do climate change, sustainability, and economic development considerations overlap in cities? *Environmental Politics*, 27 (1), 115–138. doi:10.1080/09644016.2017.1373419

Kingdon, J.W., 1995. *Agendas, alternatives, and public policies*. Glenview, IL: Longman.

Koch, M. and Lindenthal, A., 2011. Learning within the European Commission: the case of environmental integration. *Journal of European Public Policy*, 18 (7), 980–998. doi:10.1080/13501763.2011.599968

Lafferty, W. and Hovden, E., 2003. Environmental policy integration: towards an analytical framework. *Environmental Politics*, 12 (3), 37–41. doi:10.1080/09644010412331308254

Lenschow, A., 2002. *Environmental policy integration. Greening sectoral policies in Europe*. London: Earthscan.

Lindenthal, A., 2014. Aviation and climate protection: EU leadership within the International Civil Aviation Organization. *Environmental Politics*, 23 (6), 1064–1081. doi:10.1080/09644016.2014.913873

Lodge, M. and Matus, K., 2014. Science, badgers, politics: advocacy coalitions and policy change in bovine tuberculosis policy in Britain. *Policy Studies Journal*, 42 (3), 367–390. doi:10.1111/psj.2014.42.issue-3

Mintrom, M. and Norman, P., 2009. Policy entrepreneurship and policy change. *The Policy Studies Journal*, 37 (4), 649–667. doi:10.1111/psj.2009.37.issue-4

Nilsson, M. and Eckerberg, K., eds., 2007. *Environmental policy integration in practice. Shaping institutions for learning*. London: Earthscan.

Nohrstedt, D., 2011. Shifting resources and venues producing policy change in contested subsystems: a case study of Swedish signals intelligence policy. *Policy Studies Journal*, 39 (3), 461–484. doi:10.1111/psj.2011.39.issue-3

Palmer, J., 2015. How do policy entrepreneurs influence policy change? Framing and boundary work in EU transport biofuels policy. *Environmental Politics*, 24 (2), 270–287. doi:10.1080/09644016.2015.976465

Remling, E., 2018. Depoliticizing adaptation: a critical analysis of EU climate adaptation policy. *Environmental Politics*, 27 (3), 477–497. doi:10.1080/09644016.2018.1429207

Rietig, K., 2013. Sustainable climate policy integration in the European Union. *Environmental Policy and Governance*, 23 (5), 297–310. doi:10.1002/eet.1616

Rietig, K., 2018. The link between contested knowledge, beliefs, and learning in European climate governance: from consensus to conflict in reforming biofuels policy. *Policy Studies Journal*, 46 (1), 137–159. doi:10.1111/psj.v46.1

Rietig, K. and Perkins, R., 2018. Does learning matter for policy outcomes? The case of integrating climate finance into the EU budget. *Journal of European Public Policy*, 25 (4), 487–505. doi:10.1080/13501763.2016.1270345

Runhaar, J., Driessen, P., and Uittenbroek, C., 2014. Towards a systematic framework for the analysis of environmental policy integration. *Environmental Policy and Governance*, 24 (4), 233–246. doi:10.1002/eet.1647

Schoenefeld, J. and Jordan, A., 2019. Environmental policy evaluation in the EU: between learning, accountability and political opportunities? *Environmental Politics*, 28 (2). [This issue].

Searchinger, T., et al., 2008. Use of U.S. croplands for biofuels increases greenhouse gases through emissions from land use change. *Science*, 319 (5867), 1238–1240. doi: 10.1126/science.1151861

Sharman, A. and Holmes, J., 2010. Evidence-based policy or policy-based evidence gathering? Biofuels, the EU and the 10% target. *Environmental Policy and Governance*, 20 (5), 309–321. doi:10.1002/eet.543

Skjærseth, J.B. and Wettestad, J., 2010. Making the EU emissions trading system: the European Commission as an entrepreneurial epistemic leader. *Global Environmental Change*, 20 (2), 314–321. doi:10.1016/j.gloenvcha.2009.12.005

Skovgaard, J., 2014. EU climate policy after the crisis. *Environmental Politics*, 23 (1), 1–17. doi:10.1080/09644016.2013.818304

Slominski, P., 2016. Energy and climate policy: does the competitiveness narrative prevail in times of crisis? *Journal of European Integration*, 38 (3), 343–357. doi:10.1080/07036337.2016.1140759

Steinebach, Y. and Knill, C., 2017. Still an entrepreneur? The changing role of the European Commission in EU environmental policy-making. *Journal of European Public Policy*, 24 (3), 429–446. doi:10.1080/13501763.2016.1149207

Toke, D., 2008. The EU renewables directive: what is the fuss about trading? *Energy Policy*, 36 (8), 3001–3008. doi:10.1016/j.enpol.2008.04.008

Weible, C.M. and Sabatier, P.A., 2009. Coalitions, science, and belief change: comparing adversarial and collaborative policy subsystems. *Policy Studies Journal*, 37 (2), 195–212. doi:10.1111/psj.2009.37.issue-2

Wurzel, R. and Connelly, S., eds., 2011. *The European Union as a leader in international climate change politics*. London: Routledge.

Wurzel, R., Liefferink, D., and Di Lullo, M., 2019. The Council, European Council and member states: changing environmental leadership dynamics in the European Union. *Environmental Politics*, 28 (2). [This issue].

Zito, A., Burns, C., and Lenschow, A., 2019. Special issue introduction: is the trajectory of European Union environmental policy less certain?. *Environmental Politics*, 28 (2). [This issue].

The European Council, the Council and the Member States: changing environmental leadership dynamics in the European Union

Rüdiger K.W. Wurzel [ID], Duncan Liefferink [ID] and Maurizio Di Lullo [ID]

ABSTRACT
The leadership dynamics between the European Council, the Council and the Member States in European Union (EU) environmental policy since the 1970s are analysed. The puzzle is that, although the EU was set up as a 'leaderless Europe', it is widely seen as an environmental leader, albeit sometimes as a one-eyed leader amongst the blind. While differentiating between leadership types, it is argued that the European Council has the largest structural, the Council the most significant entrepreneurial, and the Member States the most important cognitive and exemplary leadership capacities. Most day-to-day environmental policy measures are negotiated by the Environment Council (in collaboration with the European Parliament). The European Council's increased interest in high politics climate change issues is largely due to the EU's global leadership ambitions. Member States have traditionally formed environmental leadership alliances on an ad hoc basis although this may be changing.

Introduction

Here, we assess the leadership dynamics between European Union (EU) Member States, the Council of the EU (Council for short) and the European Council in EU environmental policy since the early 1970s. Scholars have paid surprisingly little attention to the Council and European Council – especially considering their central importance for EU (environmental) policymaking and European integration. Even the seminal *Environmental Politics* special issue, 'A green dimension for the European Community: Political issues and processes' (Judge 1992), lacked an article focusing specifically on any of these actors.

Although the European Council and Council are central actors in EU (environmental) policymaking, they share decision-making powers with other EU, Member State and societal actors. After the Second World War, the founding

Member States deliberately set up the EU as a 'leaderless Europe' in which decision-making powers are spread among a relatively wide range of actors resulting in 'the European Union deliberately shunning the institution of an overriding leadership' (Hayward 2008, p. 1). At first sight, the EU therefore seems ill-equipped to offer leadership. Nevertheless, scholars have widely portrayed the EU as an environmental leader, albeit sometimes as a one-eyed leader amongst the blind (e.g. Oberthür and Roche Kelly 2010). This creates a puzzle, which we aim to resolve by focusing on the European Council, Council and Member States' abilities to provide leadership. Our contribution answers the following two main research questions: First, which types of leadership have the European Council, Council and Member States offered in EU environmental policy? Second, how have the leadership dynamics developed between these core EU environmental policy actors since the early 1970s?

We argue that EU environmental policy and EU integration are inextricably linked. In the intellectual tug of war between intergovernmentalists and neofunctionalists that initially dominated EU studies, the analytical focus was on whether Member States (intergovernmentalists) or supranational EU institutions (neofunctionalists) dominate the EU policymaking process and European integration. In this important scholarly debate the focus was primarily on *who* provides leadership rather than on *what type of leadership* core actors offer. We start with an explanation of the different leadership types, which we then use to analyse the changing roles of the European Council, Council and Member States. In conclusion, we assess the changing leadership dynamics and offer a critical reassessment of our puzzle and the main research questions.

Leaders and leadership

International relations (IR) scholars (e.g. Young 1991) first recognised the ability of environmental leader states to act as drivers of change before it gained traction in comparative politics and EU policy studies (e.g. Andersen and Liefferink 1997, Liefferink and Andersen 1998). While drawing especially on Burns (1978) and Young (1991), Liefferink and Wurzel (2017) and Wurzel et al. (2017) differentiated between four *types* of leadership: structural, entrepreneurial, cognitive and exemplary. Here, we draw on these four analytical leadership types while linking them to EU integration theories such as 'old' and 'new' intergovernmentalist approaches (e.g. Hoffmann 1966, Bickerton et al. 2015) and 'old' and 'new' neofunctionalist theories (e.g. Haas 1958, Sandholtz and Stone Sweet 1998).

We argue that *structural* leadership largely follows the intergovernmentalist logic according to which (the most powerful) Member State actors dominate EU policymaking and European integration. *Entrepreneurial* leadership is broadly compatible with neofunctionalist reasoning which emphasises the importance of functional cooperation as a means of achieving compromises or, as Young (1999, p. 293) put it, 'negotiating skill to frame issues in ways that foster integrative bargaining and to put together deals'. *Cognitive* leadership closely relates to constructivist approaches (e.g. Hajer 1995, Risse 2009) that emphasise the central importance of ideas for the definition of actors' interests and preferences. Finally, *exemplary* leadership bears close resemblance to policy transfer and diffusion approaches (e.g. Tews *et al.* 2003), which both assume that good examples are followed elsewhere. The analytical overlap between our four leadership types – structural, entrepreneurial, cognitive and exemplary – and four widely used EU studies theories – intergovernmentalist, neofunctionalist, constructivist and policy transfer theories – is not perfect. However, this does not diminish our core argument that a multifaceted leadership concept, which cuts across well-established, rival theories of EU integration and/or policies, can provide novel analytical insights.

First, *structural* leadership is widely associated with military power, which plays no significant role in resolving EU environmental problems. Importantly, we can also link structural leadership to economic power, e.g. the EU's Single European Market and formal institutional powers (Wurzel *et al.* 2017, p. 289). Power is important for actors who want to exert structural leadership. However, although '[a]ll leaders are actual or potential power holders, ... not all power holders are leaders' (Burns 1978, p. 19). According to Young (1991, p. 288) 'structural leaders are experts in translating the possession of material resources into bargaining leverage'. Especially, intergovernmentalists view the European Council, bringing together the Member States at the highest political level, as the most powerful EU institution (Hoffmann 1966, Puetter 2014, Bickerton *et al.* 2015).

Second, *entrepreneurial* leadership involves the use of diplomatic and negotiating resources, which are needed to broker compromise agreements that offer all parties benefits. An entrepreneurial leader is often 'an agenda setter and popularizer who uses negotiating skill to devise attractive formulas and to broker interests' (Young 1991, p. 300). According to neo-functionalists, Member State officials, interest groups and supranational institutions play a key role in fostering joint solutions to common problems (Haas 1958, Sandholtz and Stone Sweet 1998). As we discuss below, the rotating six-monthly EU Presidency can act as agenda-shaper and facilitate compromise solutions (Tallberg 2006). Importantly, we do not count actions by environmental laggards who try to water down or prevent the adoption of EU environmental measures as entrepreneurial leadership. As Underdal (1994, pp. 178–9) has argued: 'leadership is associated with the collective pursuit of some common good or joint purpose'.

Third, *cognitive* leadership requires the generation and provision of ideas and expertise that can lead to the re-/definition of actors' interests and preferences (Young 1991, Hajer 1995, Risse 2009). Examples include concepts such as: ecological modernisation, which assumes that ambitious environmental measures are beneficial for both the environment and economy; and the low carbon economy, which aims to reduce greenhouse gas emissions (GHGE) while creating jobs in e.g. the renewables industry (Wurzel *et al.* 2017). Importantly, scientific expertise and experiential knowledge about how new policies or instruments (e.g. emissions trading schemes (ETS)) actually work also constitute cognitive leadership resources (Haverland and Liefferink 2012).

While large Member States tend to have greater structural leadership capabilities than small Member States, the same does not necessarily apply to cognitive environmental leadership capabilities. Denmark, Sweden and the Netherlands have consistently provided important cognitive leadership for EU environmental policy (e.g. Andersen and Liefferink 1997) while, for instance, Belgium has periodically supplied cognitive leadership on climate change issues (Interview, Member State officials, 2016–2017). Liefferink and Wurzel (2017) have argued that in terms of cognitive leadership small Member States may punch well above their structural leadership weight.

Fourth, *exemplary* leadership (or leadership by example) implies the *intentional* or *unintentional* setting of good examples. *Intentional* exemplary leadership entails the unilateral adoption of ambitious domestic environmental measures, which aim to attract followers. It resembles Grubb and Gupta's (2000) notion of directional leadership. *Unintentional* exemplary leaders (or pioneers, see Liefferink and Wurzel 2017), however, do not usually try to attract followers although unintentionally they may nevertheless offer models for others. The policy transfer and learning literature contains intentional *and* unintentional examples that other actors emulate (e.g. Tews *et al.* 2003).

Different leadership types are usually combined. For example, an actor may facilitate coalition-building (entrepreneurial leadership), provide scientific expertise (cognitive leadership) and set an example (exemplary leadership). According to the state-centred leadership literature, the specific mix of different leadership types that particular actors employ varies across issues and over time (Liefferink and Wurzel 2017). Usually more than one type of leadership is necessary to achieve integrative institutional bargaining success in environmental policy (Young 1991). In other words, structural leadership on its own will not always win the day. Here, we assess whether this applies also to the European Council and Council.

European Council

For new intergovernmentalists, 'the European Council has established itself as a pivotal institutional actor' (Puetter 2015, p. 165). Dupont and Oberthür (2017, p. 66) have argued that the European Council and Council 'are simultaneously meeting places for Member States (at Ministerial level in the Council and at the level of heads of state or government in the European Council), and [EU] institutions in their own right, which decide on internal and external [EU] negotiation positions' (similarly, see Hayes-Renshaw and Wallace 2006).

Composition

The European Council consists of Member States' most senior political representatives, i.e. the Heads of State or Government. The President of the European Commission also attends their meetings while the Foreign Ministers did so until 2009. Prior to the 2009 Lisbon Treaty, the six-monthly rotating EU Presidency chaired all European Council and Council meetings. Since 2009, the European Council has a President who is elected through qualified majority voting (QMV) by the Heads of State or Government. While the European Council's composition has varied over time, it has always been only the Heads of State or Government who take decisions, normally by consensus.

Types of leadership

Although observers widely see the European Council as the most senior and powerful EU institution (Hayes-Renshaw and Wallace 2006, Puetter 2014), it does not have law-making competences. Instead, it is the Council that, together with the European Parliament (EP), adopts EU laws. According to the 2009 Lisbon Treaty (article 15), the 'European Council shall provide the Union with the necessary impetus for its development and shall define the general political directions and priorities thereof. It shall not exercise legislative functions'.

The European Council became a formal EU institution only with the 2009 Lisbon Treaty. However, already before 2009, the Heads of State or Government acted 'as arbiters of last resort in Council decision-making, as an informal but continuous body, governed not by the treaties but by its own rules of procedure' (Janning 2005, p. 825). Meetings by the Heads of State or Government were initially referred to as summits; they became institutionalised only in 1974 following a Franco-German initiative. However, while the 'Franco-German duumvirate' (Paterson 2012) has, in line with intergovernmentalist logic, long played a central role for the deepening of European integration, it has remained inconsequential for EU environmental policy because France and Germany have different environmental priorities, instruments and regulatory styles (Héritier et al. 1996).

Its *internal* and *external* powers provide the European Council with considerable *structural* leadership capabilities which, however, it has only periodically activated for environmental policy. Examples include the European Council's involvement shortly before and/or after important international environmental meetings. For example, in October 1972, a European summit gave the green light for a common EU environmental policy after the June 1972 UN Stockholm conference had exposed its absence (Bungarten 1978, Judge 1992). It was only the 1987 Single European Act (SEA) that introduced explicit Treaty-based EU environmental policy competences. However, the European Council did not restrain the Council from adopting (in consultation with the EP) a large number of EU environmental laws before 1987 (see Figure 1). Other Treaty provisions, e.g. the harmonisation of Member State laws to create a common market, offered the legal basis for it. Over the years, the European Council endorsed the strengthening of environmental provisions in, for example, the 1997 Amsterdam Treaty, which stipulated the principles of sustainable development and environmental policy integration (EPI). By supporting the elevation to Treaty level of such action-guiding norms, the European Council merely played catch-up with the Environment Council which had already accepted them.

By mid-2018, the European Council had not acted as a supreme arbiter for unresolved disagreements on environmental dossiers in the Council with the exception of some climate change dossiers (Interviews, 2017–2018). During the 2020 climate and energy package negotiations, insurmountable differences emerged between the poorer ('new') Central and Eastern European States (CEES) and the richer ('old') Western European Member States (Interviews, EU and Member States officials, 2016–2017). Functional integration and problem-solving had reached its limits at the Council level. With the legislative co-decision procedure ongoing between the Council and the EP, the European Council agreed detailed compromise solutions in December 2008. The European Council asked the Council to integrate its proposed compromises in negotiations with the EP. The European Council, being the most senior EU institution, thus used its structural leadership capacity by, on this high politics issue, *de facto* taking away the Council's prerogative to conduct legislative negotiations with the EP.

For the 2015 UN Paris Climate Conference, all Parties had to submit their future emission reduction plans – Intended Nationally Determined Contributions (INDCs) – in 2014. Since these plans encompassed elements cutting across several policy areas and again pitted the CEES against the Western European Member States, the European Council decided to take a stance in October 2014 while producing a detailed plan (with a 2030 time horizon) to be presented in Paris (Council 2008, 2009, p. 8). The European Council's structural leadership capacity enabled it to draft this plan, which contained elements that would later form the basis of the Commission's legislative proposals. Although traditionally the Commission jealously guards its *formal* right of initiative, it broadly accepted the European Council's plan. This enabled the Commission to submit proposals, confident of the plan's acceptability to the Council since the European Council had already reached agreement on the headline targets and basic principles.

These examples provide empirical evidence for new intergovernmentalists' claims that the European Council is becoming more involved in detailed policy-making thus triggering 'integration without supranationalism' (Bickerton *et al.* 2015). One senior Member State official (Interview, 2017) cautioned that the European Council's regular involvement in (high politics) climate change issues is unlikely to continue because they [i.e. the Heads of State or Government] have other political issues to deal with, including Brexit. Second, what they have done in the October 2015 deal is probably the level of specificity which they could have done and it becomes more technical on emissions trading and other issues… And you need quite a long time to prepare for that.

However, when the EU revises upwards the 2030 climate change targets and establishes the 2040 targets, the European Council is likely to be involved again. As long as significant differences in GDP and/or energy mixes between Member States remain, the European Council may well be called upon again to mediate conflicts that have remained unresolved at Council level.

EU environmental/climate policy often has a highly technical, scientific character, and thus requires cognitive leadership capacity that is more readily available in the Council than the European Council. The so-called *Leaders' Agenda* (European Council 2017, p. 2), which the European Council adopted in October 2017 to identify core agenda items for its meetings between October 2017 and March 2019, merely listed climate change (as the only environmental issue) for possible discussions at one of its forthcoming meetings. However, there is an 'agreement that each and every Head of State or Government, if he or she feels left behind, could call a follow-up' (Interview, Member State official, 2017). Accordingly, Member States can request a follow-up during European Council meetings of issues that the Council had already decided. On climate and energy issues, Member States have derived this 'procedural right' from the Conclusions of the European Council (2014, p. 1) meeting of October 2014 which state: 'The European Council will keep all the elements of the framework under review and will continue to give strategic orientations as appropriate, notably in respect to consensus on ETS, non-ETS, interconnections and energy efficiency'. This statement was itself the result of an informal promise that no Member State should be left behind, which the French President, Nicolas Sarkozy, had made during the 2020 climate and energy package negotiations. However, by mid-2018 this procedure was rarely used with only the Polish government calling for a European Council follow-up regarding the revision of the EU ETS by the Council (and EP) in late 2017.

The European Council's *entrepreneurial* leadership became apparent, for example, at its 2003 Thessaloniki meeting which set up the Green Diplomacy Network (GDN) in order to strengthen the EU's foreign environmental policy (Council 2003). The GDN and European External Action Service (EEAS), which became operational in 2003 and 2011, respectively, have increased the EU's entrepreneurial leadership capacity. The same applies to the increase in the number of the European Council's meetings since the 2000s and the elected President who the Council Secretariat supports in his/her task to prepare meetings; these developments ensure continuity and facilitate consensus within the European Council (Council Secretariat 2017). However, for environmental policy the European Council cannot usually match the Council's entrepreneurial leadership capacities.

The European Council has rarely offered *cognitive leadership* on environmental issues although it has regularly done so on general European integration issues. One exception constitutes the European Council's endorsement of the so-called Cardiff Strategy, which demanded better integration of environmental concerns by all Council formations; a meeting in Cardiff under the 1998 UK EU Presidency adopted the strategy, which soon ran out of steam (Wurzel 2004).

For the international climate change negotiations, the European Council has frequently endorsed the EU's global leadership ambitions through *exemplary leadership*. This has manifested itself in, for example, the adoption of relatively ambitious GHGE reduction goals and renewable energy targets, the details of which the Council and the EP normally negotiate following a formal Commission proposal.

Council of the European Union

Legally speaking, there is only one Council of the European Union – referred to as Council of Ministers until 2009 – although the ministers (and their officials) responsible for particular policy areas usually meet separately in different Council formations such as the Agricultural Council and Environment Council, the latter of which is this section's main focus.

Composition

The ministerial meetings constitute only the tip of the iceberg of the Council machinery, which includes also the Committee of Permanent Representatives (Coreper) and the Council Working Groups (e.g. Hayes-Renshaw and Wallace 2006). As predicted by neofunctionalist theories (e.g. Haas 1958, Sandholtz and Stone Sweet 1998), over time European integration 'spilled over' into new policy areas including environmental policy. By the 1990s, the number of different Council formations increased to more than 20. The European Council curtailed them at ten in 2009. The growing importance of the Environment Council is evident from its increased number of annual meetings, which rose from one (1973–1982), to two (1982–1989) to a minimum of three (since 1989). However, the increasingly important practice of informal trilogues between the Commission, Council and EP, which aim to speed up decision-making, may over time lead to a reduction in Environment Council meetings. In trilogues the Environment Council has no formal role because the Council negotiation team is led by Presidency officials while the Council's mandate is agreed at Coreper level. The 2017 Estonian Presidency was the first in many years to organise only one Environment Council meeting as informal trilogues dealt with many of its dossiers.

Coreper and Council Working Groups prepare the ministerial meetings, which national officials from the Brussels-based Permanent Representations attend, frequently joined by officials from national Ministries. For neofunctionalists, this type of *engrenage* (getting caught in the gears) between EU institutional and Member State actors is in line with the Monnet method which aims to bring about deeper political European integration through functional cooperation or 'integration by stealth' (Hayward 2008). Seen from a leadership type perspective, the multi-layered Council machinery with its deep reach into Member State bureaucracies offers significant opportunities for entrepreneurial leadership and, to a lesser degree, cognitive leadership.

Since the creation of the Environment Council in 1973, the Environment Working Group has held three to four meetings weekly. As the international climate change negotiations advanced and the workload on climate-related issues increased significantly for the Council, it established a Working Party (WP) on climate-related issues separate from the Environment Working Group in the 2000s. In 2001, the Council renamed it the WP for International Environmental Issues with several sub-formations including climate change. Since 2001, there have been two WPs, one on internal and one on international environmental issues (Council 1999, 2001). Four expert groups – on further action, mitigation, adaptation and implementation – undertook the preparatory work for the WP for international environment issues (climate change), which the Council reorganised following the 2015 Paris climate conference. In doing so, the Council further increased its entrepreneurial and cognitive leadership capacities.

Until the 2009 Lisbon Treaty, the six-monthly rotating EU Presidency was responsible for chairing all meetings of both the Council and European Council. The 2009 Lisbon Treaty retained the rotating Council Presidency for almost all Council formations but created an elected European Council President. The rotating Presidency must fulfil the following, at times conflicting, five main roles which require, in particular, entrepreneurial leadership: manager and administrator, honest broker, initiator, point of contact (for other EU institutions and Member States), and representative in international negotiations. While the Council Secretariat tends to emphasise the honest broker role, some large Member States (in particular, France and the UK) have stressed the initiator role (Wurzel 2004).

Since the 1980s, most Presidencies have also organised one informal Environment Council meeting. Such meetings, aiming at encouraging frank exchanges, have no formal agenda (Council 2015). They usually discuss broad themes (e.g. ecological industrial policy) that the incumbent Presidency proposes rather than EU legislation and thus provide opportunities for cognitive leadership.

Types of leadership

Compared to the European Council, the Council has generally higher entrepreneurial environmental leadership capacities partly due to the existence of Coreper and Council Working Groups which Member State officials attend. Following the neo-functionalist logic, this type of *engrenage* between EU and national institutional actors can facilitate joint solutions (Haas 1958, Sandholtz and Stone Sweet 1998). The General Secretariat of the Council (Council Secretariat) supports the Council; it is responsible for the organisation of all Council meetings in Brussels, ensures that rules and procedures are followed and acts as a confidential advisor behind the scenes. Presidencies by smaller Member States tend to rely more heavily on the Council Secretariat than large Member States, not least because they have smaller entrepreneurial leadership capacities (e.g. ministerial staff) to cover the wide range of often highly technical EU environmental policy issues.

Prior to the 2009 Lisbon Treaty, the EU reformed the rotating Presidency only incrementally. Arguably, the most important reform was the introduction of the so-called trio Presidency with at least one large Member State forming part of a 'team' of three Member States. This reform bolstered both the entrepreneurial and structural leadership capacities of the Council. As the Presidency (together with the Commission and the EEAS) represents the EU externally, it is seen as beneficial for the EU's interests that trio Presidencies are able to draw on the diplomatic resources and structural powers of large Member States in international negotiations (Interviews, EU and Member State officials, 2014–2017). For the climate policy issue, leaders and lead negotiators were created in the 2000s to allow for continuity beyond the rotating EU Presidencies (Dupont and Oberthür 2017). This reform further contributed to the external leadership capacities of the Council. France, Germany and the UK (i.e. three of the four large Member States) as well as the Commission have usually held the most important among those positions (Interviews, 2013–2017).

If one accepts that authoritative decision-making in the form of the adoption of legally binding acts amounts to structural leadership, then the Council also has (together with the EP) considerable *structural* leadership powers in EU environmental policy. With a few exceptions in the field of climate and energy policy where the European Council took the lead (see above), the Environment Council is still the arena where Member States negotiate and assert their powers on environmental matters. According to new intergovernmentalists (e.g. Bickerton *et al.* 2015), there has been a decline in the Council's overall legislative output in the post-Maastricht era (i.e. after 1992). Figure 1 shows a significant drop in the Council's adoption rate of legally binding environmental acts although only since 2009 and with the exception of 2012–2014.

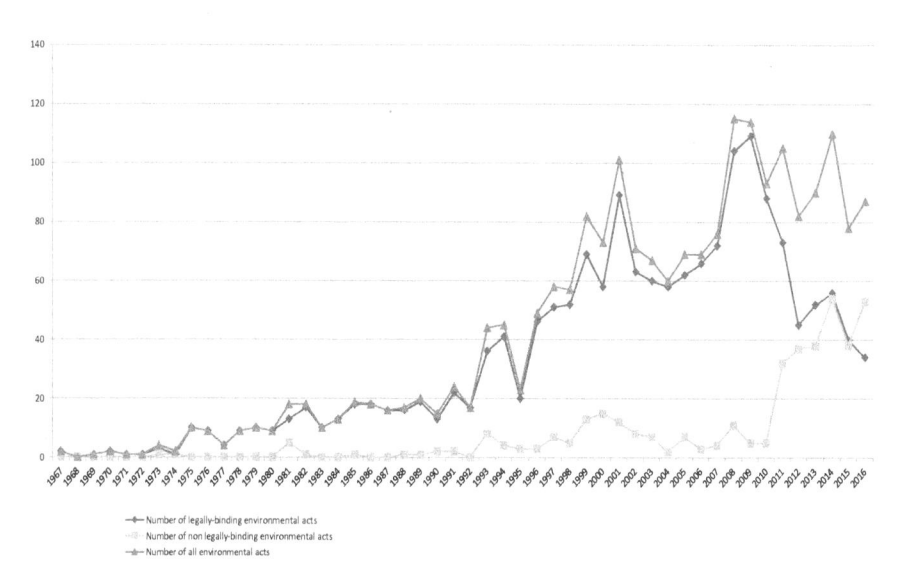

Figure 1. Environmental acts 1967–2016.

Note: Legally binding environmental acts include directives, regulations, decisions and comitology decisions. Non-legally binding environmental acts refer to non-legislative acts including opinions and recommendations. Source: Eur-lex 2017 and Council Secretariat (2017).

Bauer and Knill (2014) have differentiated policy measures according to their *density* (i.e. number of measures adopted) and *intensity* (i.e. relative importance of measures). Figure 1 illustrates only the density of legally binding and non-binding measures. Due to word constraints, we focus on adoption trends (policy density).

Figure 1 shows that the 1987 SEA, which introduced explicit environmental Treaty provisions, had no discernible effect on the adoption rate of legally binding environmental acts. Contrary to claims by new intergovernmentalists (Puetter 2014, Bickerton et al. 2015), the ratification crisis of the Maastricht Treaty in 1992, which resulted in the adoption of the subsidiarity principle (according to which decisions should be taken at the lowest possible level), did not trigger a decline in the adoption rate of legally binding environmental acts; on the contrary, it actually rose significantly with the exception of 1995. Yet, the Lisbon Treaty's entry into force in 2009 seems to have triggered a significant downturn in the adoption of legally binding environmental acts. However, other factors have also played an important role including the economic crisis, the Commission's better regulation agenda and its Regulatory Fitness and Performance (REFIT) programme and the maturation of EU environmental policy (Interviews, EU officials, 2016–2017, see also Zito et al. 2019 – this volume)[1].

The Council's *cognitive* leadership capacity appears less significant than its entrepreneurial and structural leadership capacities although there are exceptions. Following the rejection of the draft EU Constitution in referendums in the Netherlands and France in 2005, Ministers represented in the Council identified the environment in general and climate change in particular as 'a new *raison d'être*' for the EU (e.g. Miliband 2006). The use of an environmental public discourse to gain support for the EU chimes well with cognitive leadership accounts and constructivist accounts (Hajer 1995, Risse 2009). However, not all Member States have been equally convinced of the need for ambitious EU environmental policies and/or deeper European integration. Following the EU's Eastern enlargements in the 2000s, a cognitive East–West divide has opened up on EU environmental policy issues, which changed significantly the actor dynamics in the Council (and European Council). While the 1995 enlargement of the EU by Austria, Finland and Sweden strengthened the Council's environmental credentials, the opposite seems to have happened after the Eastern enlargement in the 2000s. Especially, the poorer CEES have often perceived ambitious EU environmental policy measures as a threat to their economic development (Skjærseth 2018). As Braun (2014, p. 457) has pointed out, in the CEES '[t]here is a general disbelief in the possibility of the [EU's] climate change policy being an opportunity for business and for jobs'.

Functionally differentiated Council formations avoid grand-scale zero-sum games where the winner takes all. However, sectoral differentiation can lead to disjointed decision-making, which is unable to take into account the holistic requirements of cross-cutting policies such as environmental policy (Wurzel 2004). From a cognitive leadership perspective there is therefore a tension between the neofunctionalist EU integration logic, which favours different Council formations along functional lines that help to avoid politically divisive conflicts, and an EPI logic, which enables the integration of environmental requirements by Council formations other than the Environment Council. EPI efforts in the form of Joint Councils in which the Environment Council met, for example, with the Energy Council, flourished briefly in the 1990s. The UK's 1992 EU Presidency launched the Cardiff strategy according to which all non-Environment Council formations had to assess how they could integrate environmental requirements into their dossiers. However, by the early 2010s, the Cardiff strategy was 'as dead as a dodo' (Interview, UK official, 2012). So far, the functionalist integration logic has largely triumphed over the EPI logic.

Broadly speaking, the Council has endorsed *exemplary* environmental leadership more often than the European Council but less often than some environmental leader states (or the EP). For example, in 2008, the Council and EP adopted an EU law, which expanded the EU ETS to the aviation sector including non-EU airlines. However, following fierce lobbying of Member States – particularly France, Germany and the UK – by the USA and China, the EU put it on ice (Wurzel *et al.* 2017, p. 288).

Member States

Compared to the smaller Member States, the large EU Member States have greater *structural* power and thus potentially also greater structural environmental leadership capacities. While France, Germany, Italy and the UK participate directly in G7 and G20 meetings, which started to discuss environmental issues more regularly in the 2000s, the smaller Member States receive representation only indirectly through the President of the European Council and the Commission President who defend the EU's collective interests. However, only Germany and the UK have regularly pushed environmental issues – most of all climate issues – at G7/G20 meetings.

Permanent environmental leader coalitions have traditionally not existed at EU level. Instead, coalitions between Member States 'have to be formed on an issue-by-issue basis and remain liable to defection' (Liefferink and Andersen 1998, p. 262). Officials widely see flexible alliances between Member States as more easily facilitating compromise solutions in the Council than permanent or semi-permanent coalitions (Interviews, EU and Member States officials, 2015–2017). However, the cognitive East–West divide on environmental issues, which emerged after the EU enlargements in the 2000s, is arguably starting to have an impact on alliance building.

Member State alliances

Following Denmark's 1973 EU accession, most observers identified a 'green trio': Denmark, Germany and the Netherlands (e.g. Andersen and Liefferink 1997). The green trio extended to a 'green sextet' when Austria, Finland and Sweden joined the EU in 1995 (Liefferink and Andersen 1998). However, the EU's environmental leader states exhibit different national environmental regulatory styles, instruments and strategies (Héritier *et al.* 1996), which helps explain why neither the 'green trio' nor the 'green sextet' ever developed into a semi-permanent, let alone permanent, alliance. Nevertheless, within the Council these 'green' Member States have worked closely on specific environmental issues. In terms of structural leadership, these three/six Member States did not muster the necessary votes for QMV decisions. Therefore, we cannot explain their significant influence on EU environmental policy merely with reference to structural leadership, which would be in line with new intergovernmentalist explanations (e.g. Bickerton *et al.* 2015). In other words, structural leadership does not necessarily always trump other types of leadership. Instead, the 'green' Member States largely facilitated the adoption of EU environmental policies at a relatively high level of environmental protection through *entrepreneurial* leadership and especially by providing ideas, expertise and 'good examples', i.e. *cognitive* and *exemplary* leadership (Liefferink and Andersen 1998). Well-developed national knowledge infrastructures (e.g. research institutes) and relevant expertise facilitated their efforts (Haverland and Liefferink 2012).

In several cases, the 'green' Member States and/or other Member States provided exemplary leadership by setting 'good examples' which stimulated policy transfer. Frustrated by the Council's inability to adopt the Commission's 1992 proposal for a carbon dioxide/energy tax (due to the UK's veto of supranational taxes on sovereignty grounds), a group of like-minded countries held meetings between officials and Ministers from Environment and Finance Ministries between 1994 and 1998 who discussed the design and effect of national eco-taxes. Belgium, Denmark, Finland, France, Germany, Italy and Sweden participated in the last meetings of this informal group. The UK, which around that time adopted a significant number of national eco-taxes, also attended some meetings (Wurzel *et al.* 2013, p. 167).

Since the 1990s, environmental leader coalitions have frequently excluded one or several 'green sextet' members while including other Member States such as the UK, especially on climate issues. Following the EU's Eastern enlargements in the 2000s, a relatively stark East–West divide has emerged on EU environmental issues in general and climate change issues in particular. The CEES' comparatively low GDP levels, specific energy mixes (e.g. Poland's high dependency on coal and reliance on Russian gas) and scepticism towards concepts such as ecological modernisation and the low carbon economy help to explain this divide (e.g. Braun 2014, Skjærseth 2018). The CEES seem to have broken with the long established informal tradition that permanent alliances between Member States should be avoided in the (Environmental) Council because they can be counter-productive for finding compromise solutions (Interviews, Member State officials, 2016–2017). The Visegrad Group – Hungary, Poland, Czech Republic and Slovakia – is a relatively small, homogenous alliance with regular meetings chaired by fixed-term presidencies. It has held meetings on climate and energy issues with the aim of defending the group's interests at the EU level. The Visegrad Group wants to progress more slowly towards full decarbonisation than other Member States, which ought to take on a bigger share in the EU's collective GHGE reduction targets. Over time, the Visegrad Group tried to expand its reach to other countries (including Bulgaria and Romania) and invited newly acceded Croatia as an observer; this, however, has made the group less homogenous.

Partly in reaction to the Visegrad Group's climate activities, the UK initiated the Green Growth Group (GGG) in 2016 (Interviews, EU and Member State officials, 2015-2017). The GGG is a fairly large, loose alliance with a small secretariat.[2] As one Member State official explained (Interview, 2017), the 'Visegrad [group] is much more institutionalised, that is clear. Why is the Green Growth Group not more institutionalised? Because there is a fine line between leadership by a group of countries ... and getting everybody on board'. The GGG, which promotes ambitious climate and energy targets while arguing that they can promote economic growth, stages annual ministerial meetings, stakeholder meetings and thematic workshops. The GGG's ministers also hold informal meetings in the margins of the Environment Council where, however, they do not act on behalf of the group. The GGG has provided cognitive and exemplary leadership by, for example, showcasing existing national 'green growth' measures and by promoting more ambitious supranational and international climate targets. The GGG has tried to enlist the help of the EP, thus also exhibiting entrepreneurial leadership.

In 2003, the Member States set up the Green Development Network (GDN) which is meant to integrate environmental objectives into the EU's foreign policy. On climate change, the Foreign Ministries of Germany and the UK as well as France engaged in coordinated outreach activities in the run up to the 2015 UN Paris climate conference (Interviews, 2015-2017). The EEAS did not coordinate these activities, which rely mainly on cognitive and exemplary leadership; EEAS has tried - with various levels of success - to 'keep the flock of 28 EU sheep together' (Interview, EEAS official, 2013) on EU foreign environmental policy issues. The EEAS did however coordinate a Climate Diplomacy Day with outreach climate-related activities by its staff in about 60 countries.

Conclusion: leadership types and dynamics

Table 1 provides a summary overview of our four leadership types and how the European Council, Council and Member States have *predominantly* used them. It also explains briefly the core roles that the European Council, Council and Member States have played on environmental issues at different governance levels. Here, we assess Member States' different types of leadership on the state/sub-state level only if they have a direct impact on the European Council and/or Council or on alliances between Member States.

Table 1. Environmental governance levels, core roles and leadership types.

	European Council	Council	Member States
Governance levels and core roles	• **EU:** Treaty changes; political guidelines; arbiter of last resort for Council disagreements. • **External:** High politics environmental issues.	• **EU:** Adoption of EU laws/policies (shared with EP). • **External:** Representation of EU sometimes shared with Member States and/or Commission.	• **State/sub-state:** Adoption of national/subnational laws/policies. • **EU:** Representation of Member States in Council and European Council. • **External international:** Member States usually represented in international environmental fora. Only large Member States directly represented in G7/G20.
Leadership types			
Structural leadership	*Very important for 'high politics' issues. Increased since 2000s:* • **EU:** Approval of Treaties. Occasionally final arbiter of Council disputes. • **External:** Overriding structural leadership for 'high politics' issues grown since 2000s.	*Important but usually shared with other EU institutions:* • **EU:** Environment Council's day-to-day decisions without European Council interference. • **External:** Council often shares negotiating powers with Commission and, if mixed agreements, with Member States.	*Important, but large states with higher structural leadership capacity than small states:* • **State/sub-state:** Varies between states. • **External EU:** Green trio/sextet. Little relevance of Franco-German alliance. North–South differences superseded by East–West divide. • **External international:** Particularly large Member States can act as environmental leaders internationally (e.g.G7/G20).

(*Continued*)

Table 1. (Continued).

	European Council	Council	Member States
Entrepreneurial leadership	*Important broker for large package deals*: ● **EU**: Large cross-sectoral package deals rarely include environmental issues. ● **External**: Green Diplomacy Network (2003) and European External Action Service (2010).	*Very important broker for routine, day-to-day decision-making*: ● **EU**: Council machinery (Coreper, Council Working Groups (including Member State officials) and Secretariat) strives to negotiate compromise deals. Rotating Presidency shapes agenda. ● **External**: For climate issues (trio) Presidency supported by lead negotiators and issue leaders.	*Important temporary, ad hoc alliances*: ● **State/sub-state**: Entrepreneurial environmental leadership varies between states. ● **External EU**: Ad hoc alliances of 'green' states have agenda setting/shaping capabilities on specific issues. Green trio/sextet until 1990s. Since the 2000s, Green Growth Group (vs. Visegrad). ● **External international**: Green diplomacy network and national environmental diplomatic efforts.
Cognitive leadership	*Important for EU integration and 'high politics' environmental issues*: ● **EU**: Cognitive leadership usually only for important integration issues ● **External**: Mainly 'High politics' issues (e.g. EU global climate leadership).	*Important role for channelling cognitive leadership*: ● **EU**: Occasional cognitive leadership, particularly by Informal Council meetings (e.g. Cardiff Strategy). ● **External**: EU Presidencies occasionally propagate ideas internationally (e.g. low carbon economy).	*Very important cognitive leadership although mainly by environmental leader states*: ● **State/sub-state**: Variable cognitive environmental leadership capacities. ● **External EU**: Individual Member States and 'green' trio/sextet offered cognitive leadership until the late 1990s. Since the 2000s, more issue specific alliances including Green Growth Group as loosely institutionalised alliance. ● **External international**: France, Germany, Italy and the UK could use their cognitive leadership capabilities in G7 especially when holding G7 Presidencies but only Germany and the UK have done so regularly.

(Continued)

Table 1. (Continued).

	European Council	Council	Member States
Exemplary leadership	*Primarily external exemplary leadership:* • **EU**: European Council rarely involved in EU internal environmental exemplary leadership decisions. • **External**: European Council often involved in high politics external exemplary leadership decisions.	*EU and external exemplary leadership:* • **EU**: Environment Council frequently supports internal exemplary leadership. • **External**: Environment Council always involved in external environmental exemplary leadership decisions.	*State/sub-state, EU and external exemplary leadership:* • **State/sub-state**: Exemplary leadership offered by different Member States for different issues. • **EU**: Member State policies often serve as examples for EU. Ambitious domestic leader state policies avoid credibility gap. • **External**: External leadership ambitions require national policies to avoid credibility gap.

Based on our analytical framework and empirical findings, we have argued that for EU environmental policy, broadly speaking the European Council has the largest structural leadership capacity, the Council has the most significant entrepreneurial leadership capacity and the Member States have the most important cognitive and exemplary leadership capacities (see grey shaded boxes in Table 1).

Over time, the leadership dynamics between the European Council, Council and Member States have evolved. The European Council gave the starting signal for a common environmental policy in the early 1970s (Bungarten 1978) and has taken a close interest in high politics climate change issues since the 2000s (Dupont and Oberthür 2017). Generally speaking, the Council has however dealt with day-to-day EU environmental policy decisions without much interference from the European Council. We were able to find only two examples of EU environmental policy (both of which relate to high politics climate change issues) since the early 1970s that support the claim by new intergovernmentalists (Bickerton et al. 2015) that the European Council is acting as arbiter for disagreements at Council.

New intergovernmentalists are correct in arguing that a decline in EU legislation has taken place although, contrary to their view, for EU environmental policy it did not set in in the post-Maastricht period (i.e. shortly after 1992) but has occurred only since 2008. One important reason for this is that the Council's entrepreneurial leadership capacity has enabled it (together with the EP) to carry on adopting a significant number of legally binding EU environmental laws until the wider political context changed significantly due to the economic crisis, the constitutional crisis surrounding the adoption of the 2009 Lisbon Treaty and the Commission's better regulation initiatives and REFIT programmes.

Since the EU's Eastern enlargements in the 2000s, significant differences in cognitive and exemplary leadership have emerged between the more affluent 'green' Member States and the poorer CEES. The Visegrad Group and, although to a lesser degree, the GGG have become relatively well-institutionalised alliances; they could change significantly the leadership dynamics in the Council and the European Council. The emergence of such alliances could herald a departure from the long established tradition that semi-permanent leader alliances should not form because they hinder the search for compromise solutions. It is for this reason that, so far, the GGG has purposefully avoided the further institutionalisation of the group.

For external EU environmental policy, the Council has to share negotiating powers with the Member States (and sometimes also the Commission). In any case, most international environmental agreements constitute so-called mixed agreements which both the EU and Member States sign (Vogler 1999). Attempts by the EEAS to coordinate EU and

Member State environmental foreign policies have had limited success despite the creation of the GDN in 2003. While the large Member States are directly represented in major non-environmental international settings, e.g. the G7/G20, which have started to discuss more regularly environmental issues, the smaller Member States are only indirectly represented through the President of the European Council and the Commission President. This grants greater structural leadership capacities to the larger Member States although only Germany and the UK have regularly used their G7 Presidencies to push environmental issues. However, in terms of cognitive and exemplary leadership, some of the smaller Member States have been capable of punching well above their structural leadership weight; this explains why they have been relatively successful in influencing EU environmental policy (Liefferink and Andersen 1998). Brexit is likely to lead to efforts among the remaining Member States to use other leadership types to compensate for the EU's reduced structural environmental leadership capacities resulting from Britain's exit.

The simultaneous use of different leadership types is usually required for successful environmental policymaking. This helps explain the mutual dependency between the European Council, Council and Member States in EU environmental policymaking. This dependency becomes most apparent in the EU's external environmental policy, notably in international climate negotiations. For instance, before and during the 2009 Copenhagen and 2015 Paris climate conferences, the European Council, Council and Member States combined different types of leadership, without much success in Copenhagen but with considerable success in Paris (Wurzel *et al.* 2017). EU actors have used the EU's relatively ambitious GHGE reduction and renewable energy targets (exemplary leadership), framing of climate change as both a threat to the environment *and* an opportunity for the low carbon economy (cognitive leadership), facilitation of alliances of states in favour of relatively ambitious climate policy measures (entrepreneurial leadership) and economic power (structural leadership) to accomplish a reduction of the 'credibility gap' (Dupont and Oberthür 2017) between the EU's external ambitions and its domestic actions.

Differentiating between different types of leadership, while taking into account the changing leadership dynamics between the European Council, Council and Member States, helps to resolve at least partly the puzzle that although the EU was set up as a 'leaderless Europe' observers have widely seen it as an environmental leader. We linked the different types of leadership in our fourfold leadership typology to existing EU integration theories. While a new intergovernmentalist perspective helps to explain the increased structural leadership offered by the European Council on high politics climate change issues, the neofunctionalist logic elucidates the interlocking relations (or *engrenage*) between EU institutional and Member State officials which also fit well an entrepreneurial leadership perspective. Constructivist approaches explain well Member States' cognitive leadership while we find many examples of exemplary leadership in the policy transfer literature.

Notes

1. The EP nearing the end of its legislative term and Member States' reluctance to agree to EU legislation shortly before national elections can also cause moderate, temporal fluctuations.
2. Belgium, Germany, Denmark, Spain, Estonia, Finland, France, Ireland, Italy, Luxembourg, the Netherlands, Portugal, Sweden, Slovenia, United Kingdom and Norway as well as the Commission have regularly attended GGG meetings. Austria has been recently asked to attend.

Acknowledgments

The authors are grateful to the referees and editors for their helpful comments on earlier drafts. The usual disclaimer applies. The authors conducted more than 20 interviews with EU and Member States officials in 2003–2004 and 2012–2017. The institutional affiliations of the interviewees are not stated to ensure non-attributability.

Disclosure statement

No potential conflict of interest was reported by the authors. The views expressed by Maurizio di Lullo are his personal views and not those of the Council of the EU.

Funding

Rudi Wurzel thanks the British Academy (grant SG 131240) and University of Hull for funding.

ORCID

Rüdiger K.W. Wurzel http://orcid.org/0000-0001-5873-4232
Duncan Liefferink http://orcid.org/0000-0002-3594-3274
Maurizio Di Lullo http://orcid.org/0000-0001-9508-1859

References

Andersen, M.S. and Liefferink, D., eds., 1997. *European environmental policy: the pioneers*. Manchester: Manchester University Press.
Bauer, M.W. and Knill, C., 2014. A conceptual framework for the comparative analysis of policy change. *Journal of European Public Policy*, 16 (1), 28–44.

Bickerton, C.J., Hodson, D., and Puetter, U., 2015. *The new intergovernmentalism.* Oxford: Oxford University Press.

Braun, M., 2014. EU climate norms in East-Central Europe. *Journal of Common Market Studies*, 52 (3), 445–460. doi:10.1111/jcms.2014.52.issue-3

Bungarten, H.H., 1978. *Umweltpolitik in Westeuropa.* Bonn: Europa Union Verlag.

Burns, J.M., 1978. *Leadership.* New York: Harper & Row.

Council, 1999. *List of committees and working parties.* Brussels: General Secretariat of the Council, *13406/99.*

Council, 2001. *List of committees and working parties.* Brussels: General Secretariat of the Council, *10279/01.*

Council, 2003. *Thessaloniki European Council 19 and 20 June.* Presidency conclusions. Brussels: General Secretariat of the Council, *11638/03.*

Council, 2008. *Energy and climate change – elements of the final compromise 17215/08.* Brussels: General Secretariat of the Council.

Council, 2009. *Brussels European Council.* 11 and 12 December 2008. Brussels: General Secretariat of the Council, 17271/08.

Council, 2015. *Handbook of the Presidency of the Council of the European Union.* Luxembourg: General Secretariat of the Council of the EU.

Council Secretariat, 2017. *The president's role.* Available from: http://www.consilium.europa.eu/en/european-council/president/role/ [Accessed 14 November 2017].

Dupont, C. and Oberthür, S., 2017. The Council and the European Council. *In*: R. Wurzel, J. Connelly, and D. Liefferink, eds. *The European Union in international climate change politics.* London: Routledge, 66–79.

European Council, 2014. *Conclusions. EUCO 169/14.* Brussels: General Secretariat of the Council.

European Council, 2017. *Leaders' agenda. Building a future together.* Brussels: European Council.

Grubb, M. and Gupta, J., 2000. Leadership. *In*: J. Gupta and M. Grubb, eds. *Climate change and European leadership.* Dordrecht: Kluwer, 15–24.

Haas, E.B., 1958. *The uniting of Europe.* Stanford, CA: Stanford University Press.

Hajer, M., 1995. *The politics of environmental discourse.* Oxford: Clarendon.

Haverland, M. and Liefferink, D., 2012. Member state interest articulation in the Commission phase. *Journal of European Public Policy*, 19 (2), 179–197. doi:10.1080/13501763.2011.609716

Hayes-Renshaw, F. and Wallace, H., 2006. *The council of ministers.* 2nd ed. Basingstoke: Palgrave Macmillan.

Hayward, J., ed., 2008. *Leaderless Europe.* Oxford: Oxford University Press.

Héritier, A., Knill, C., and Mingers, S., 1996. *Ringing the changes in Europe.* Berlin: de Gruyter.

Hoffmann, S., 1966. Obstinate or obsolete? The fate of the nation-state and the case of Western Europe. *Daedalus*, 68 (3), 862–915.

Janning, J., 2005. Leadership coalitions and change: the role of states in the European Union. *International Affairs*, 81 (4), 821–833. doi:10.1111/j.1468-2346.2005.00486.x

Judge, D., ed., 1992. Special issue. A green dimension for the European community: political issues and processes. *Environmental Politics*, 1 (4), 1–249. doi:10.1080/09644019208414043

Liefferink, D. and Andersen, M.S., 1998. Strategies of the 'green' member states in EU environmental policy-making. *Journal of European Public Policy*, 5 (2), 254–270. doi:10.1080/135017698343974

Liefferink, D. and Wurzel, R.K.W., 2017. Environmental leaders and pioneers: agents of change? *Journal of European Public Policy*, 24 (7), 951–968. doi:10.1080/13501763.2016.1161657

Miliband, D., 2006. Building an environmental union. *Speech in Berlin*, 19 September. Berlin: British Embassy Berlin.

Oberthür, S. and Roche Kelly, C., 2010. EU leadership in international climate policy: achievements and challenges. *The International Spectator*, 43 (2), 35–50. doi:10.1080/03932720802280594

Paterson, W., 2012. A contested Franco-German duumvirat. *In*: J. Hayward and R. Wurzel, eds. *European disunion*. Basingstoke: Palgrave Macmillan, 236–251.

Puetter, U., 2014. *The Council and the European Council*. Oxford: Oxford University Press.

Puetter, U., 2015. The European Council. *In*: C.J. Bickerton, D. Hodson, and U. Puetter, eds. *The new intergovernmentalism*. Oxford: Oxford University Press, 165–184.

Risse, T., 2009. Social constructivism and European integration. *In*: A. Wiener and T. Diez, eds. *European integration theory*. Oxford: Oxford University Press, 144–160.

Sandholtz, W. and Stone Sweet, A., 1998. *European integration and supranational governance*. Oxford: Oxford University Press.

Skjærseth, J.B., 2018. Implementing EU climate and energy policies in Poland. *Environmental Politics*, 27 (3), 498–518. doi:10.1080/09644016.2018.1429046

Tallberg, J., 2006. *Leadership and negotiation in the European Union*. Cambridge: Cambridge University Press.

Tews, K., Busch, P.-O., and Jörgens, H., 2003. The diffusion of new environmental policy instruments. *European Journal of Political Research*, 42 (4), 569–600. doi:10.1111/1475-6765.00096

Underdal, A., 1994. Leadership theory. *In*: I.W. Zartman, ed. *International multilateral negotiation*. San Francisco: Jossey-Bass, 178–197.

Vogler, J., 1999. The European Union as an actor in international environmental politics. *Environmental Politics*, 8 (3), 24–48. doi:10.1080/09644019908414478

Wurzel, R.K.W., 2004. *The EU presidency: 'Honest broker' or driving seat*. London: Anglo-German Foundation. http://www.agf.org.uk/cms/upload/pdfs/R/2004_R1258_e_eu_presidency.pdf

Wurzel, R.K.W., Connelly, J., and Liefferink, D., eds., 2017. *The European Union in international climate change politics*. London: Routledge.

Wurzel, R.K.W., Zito, A.R., and Jordan, A., 2013. *Environmental governance in Europe*. Cheltenham: Edward Elgar.

Young, O.R., 1991. Political leadership and regime formation. *International Organisation*, 45 (3), 349–375. doi:10.1017/S0020818300033117

Young, O.R., 1999. Regime effectiveness: taking stock. *In*: O.R. Young, ed. *The effectiveness of international environmental regimes*. Cambridge, MA: MIT Press, 249–279.

Zito, A.R., Burns, C., and Lenschow, A., 2019. Is the trajectory of European Union environmental policy less certain? *Environmental Politics*, 28 (2).

ə OPEN ACCESS

De-Europeanising or disengaging? EU environmental policy and Brexit

Charlotte Burns ⓘ, Viviane Gravey ⓘ, Andrew Jordan ⓘ and Anthony Zito ⓘ

ABSTRACT
The European Union (EU) has had a profound effect upon its members' environmental policy. Even in the United Kingdom (UK), the EU's most recalcitrant member state (historically labeled the 'Dirty man of Europe'), environmental policy has been Europeanised. As the UK moves to the EU's exit door it is timely to assess the utility of Europeanisation for understanding policy dynamics in the UK. Drawing upon interviews and extensive engagement with stakeholders, this article analyses the potential impact of Brexit upon environmental policy and politics. The analytical toolkit offered by de-Europeanisation is developed to identify the factors that drive and inhibit de-Europeanisation processes, thereby providing insights that may be applicable in other settings. Disengagement and policy stagnation are presented as more likely environmental outcomes of Brexit, with capacity emerging as a central explanatory variable.

Introduction

Until recently, scholars have assumed that Europeanisation and the development of the European Union's (EU's) environmental *acquis communautaire* were largely top down, uni-directional, expansionary and positive for environmental outcomes. The conglomerate of crises (Falkner 2016) that has recently beset the EU has challenged these assumptions (Zito *et al.* 2019). Since the publication of this journal's 1992 Special Issue on the EU and environmental policy (Judge 1992), academics have depicted the EU as a positive influence on United Kingdom (UK) environmental policy, which they typically describe as being Europeanised (Lowe and Ward 1998; Jordan 2002, 2004). The UK has downloaded numerous pieces of legislation[1] and the EU has become part of the daily business of UK environmental

---

This is an Open Access article distributed under the terms of the Creative Commons Attribution License (http://creativecommons.org/licenses/by/4.0/), which permits unrestricted use, distribution, and reproduction in any medium, provided the original work is properly cited.

stakeholders and policy-makers. The UK's anticipated exit from the EU ('Brexit') consequently poses important analytical and empirical challenges to the literatures on Europeanisation and EU and UK environmental policy.

The concepts of dismantling (Gravey and Jordan 2016; Steinebach and Knill 2017) and de-Europeanisation (Copeland 2016) are emerging as key analytical tools for understanding the current and likely future trajectory of EU environmental policy. Here, we build upon these terms and the emerging de-Europeanisation literature to offer an original assessment of the likely implications of Brexit for the UK and for our wider understanding of Europeanisation. At the time of writing, Brexit negotiations are ongoing, the final outcome is unknown. Despite this uncertainty and, recognising that Brexit represents an extreme example of possible de-Europeanisation, we contend that Brexit reflects a wider movement to reduce and limit the EU's influence on its Member States. Drawing upon Hogwood and Peters (1982) insight that current and past policies are likely to shape future policies, we argue that, in the short term, Brexit's immediate impact is likely to be limited, but over the longer term a wider divergence between the EU and UK is likely, with the prospect of UK environmental policy stagnating. Our work chimes with the emerging findings of the dismantling literature: we see little evidence that deliberate de-Europeanisation of environmental policy will follow Brexit, but suggest a longer process of disengagement is likely.

The following sections review the state of the art on the Europeanisation of UK environmental policy and then analyse the referendum campaign and its aftermath, to determine the variables likely to shape de-Europeanisation. Our research draws upon a review of primary and secondary documentary sources, interviews with key political actors drawn from different party backgrounds and evidence gathered from eight stakeholder workshops carried out between 2015 and 2018 involving representatives from environmental non-governmental organisations (ENGOs), civil servants and parliamentary officials from across the UK. The workshops operated under Chatham House rules so we report findings without direct attribution to the individuals concerned.

We make three principal contributions. First, theoretically we refine and extend current work on de-Europeanisation and identify key variables shaping the patterns of behaviour from which we can draw analytical generalisations that can be tested in other national and policy settings. Second, we use original interview and extensive stakeholder engagement data to provide robust evidence to underpin our analysis. Third, we offer one of the first and most extensive political analyses of the implications of Brexit for UK and EU environmental policy and politics. We thereby provide comprehensive analysis of one aspect of Brexit and use this case to explore and engage with wider, important debates in the public policy and Europeanisation literatures.

Europeanisation, de-Europeanisation and disengagement

The mushrooming of Europeanisation studies generated a 'bewilderingly large array of definitions' (Jordan and Liefferink 2004, p. 5); here we draw upon those proposed by Radaelli (2004) and Börzel and Risse (2003). Radaelli (2004) suggests that Europeanisation consists of processes of construction, diffusion and institutionalisation of 'formal and informal rules, procedures, policy paradigms, styles, "ways of doing things" and shared beliefs and norms which are first defined and consolidated in the EU policy process and then incorporated in the logic of domestic (national and subnational) discourse, political structures and public policies' (p. 3). Börzel and Risse (2003) use a threefold typology of the EU's impact upon a state's policy, politics and polity. Here 'policy' refers to the broader policy paradigm and specific policy goals and instruments; 'politics' includes the engagement of civil society, business actors, parties and the wider public with Europeanisation processes; and 'polity' centres on relations between levels of government and formal institutional structures, including administrative capacity and accountability, both within the UK and in the EU.

Early Europeanisation studies generally focused upon the EU's impact upon states, and, whilst retrenchment and resistance to the EU have long been part of the analytical debates about Europeanisation there have been relatively few examples of states actively trying to roll back EU policies at the national level. More recently, we have seen emerging debates around dismantling EU policy (Bauer *et al.* 2012; Jordan *et al.* 2013; Gravey and Jordan 2016) and studies elaborating the concept of de-Europeanisation (Aydın-Düzgit and Kaliber 2016; Copeland 2016; Raagmaa *et al.* 2014). Essentially de-Europeanisation amounts to dismantling EU policy at the domestic level, where dismantling means the 'cutting, diminution or removal of existing policy' (Jordan *et al.* 2013, p. 795). Copeland (2016) suggests that a key analytical component of de-Europeanisation is that it is *intentional* with 'the specific aim to reverse the process of Europeanisation and to prevent future uploading and down-loading in the governance process' (2016, p. 1126). Significantly, he distinguishes de-Europeanisation from disengagement; the latter involves a state retreating from active Europeanisation, maintaining the domestic processes and structures affected by Europeanisation, but not seeking to adapt them further to the EU's influence.

This conceptual distinction is useful in the Brexit context. The act of leaving the EU clearly constitutes an intention to de-Europeanise. However, it does not follow that the UK will actively dismantle the governance processes and policies established as a consequence of EU membership.

If, as a result of Brexit, the British government chooses to leave the vast majority of EU environmental (or indeed any other) policy in place, then, despite no longer being an EU member, UK policy may be disengaged rather than de-Europeanised. This description would be particularly apt in those policy sectors, such as environment, where the UK is likely to have to align with EU policy in order to trade with the EU 27, even though the opportunity to upload policy will no longer be available. Consequently, our endeavour is partly to determine the likelihood of active and deliberate attempts to reverse the Europeanisation of UK environmental policy, i.e. the active dismantling (Gravey and Jordan 2016) of the environmental *acquis* within the UK.

Copeland (2016) identifies two conditions shaping the level and extent of de-Europeanisation. First, the degree to which policy is centralised: policy decided by central government involves fewer veto players making it easier to reverse. Second, the level of domestic political support for the policy. Copeland argues that in the case of EU employment policy the levels of public knowledge and active support were limited, which resulted in less contestation when the policy was rolled back. Conversely, where there is a higher level of knowledge and popular support we should expect de-Europeanisation to prove more challenging. Synthesising Copeland's (2016) and Börzel and Risse (2003) work (see Table 1), we anticipate that, where Europeanisation processes have led to limited changes, de-Europeanisation will be easier to achieve; conversely, where Europeanisation has been more deep-seated, de-Europeanisation will be more difficult and hence disengagement more likely.

The following section provides a succinct review of the state of the art on the Europeanisation of UK environmental policy. We then explore the referendum campaign and its immediate aftermath to determine the likelihood of de-Europeanisation, and the patterns that may emerge in the light of our expectations.

Europeanising the UK's environment

Prior to the UK referendum there had been relatively few up-to-date academic analyses of UK environmental policy. Reviews of specific policies were produced, especially on climate change and energy policy following the UK's pioneering Climate Change Act (Carter and Jacobs 2014; Lorenzoni and Benson 2014; Lockwood 2013), but analyses of broader UK-EU policy dynamics were notable by their absence, which reflected the relative stability of UK environmental policy and the academic consensus that UK environmental policy had been Europeanised (Lowe and Ward 1998; Jordan 2002). Below we review what this meant in relation to policy, politics and polity.

Table 1. Expectations of de-Europeanisation.

Europeanisation	Policy	Politics	Polity	De-Europeanisation
Limited impact of the EU	Voluntary policy goals have resulted in limited or few policy changes, governance changes or cognitive shifts amongst policy-makers.	Limited awareness of policy outside government. Limited engagement of wider societal actors.	Centralized policy-making. Limited veto players. Low public awareness.	Likely and uncontested.
Stronger impact of the EU	Mandatory changes that have resulted in significant policy changes and infrastructural investment.	Mobilization of business and societal interests in favour of policy. Wider social and business expectations coalesced around EU policy. Contestation.	Devolved policy area. Multiple veto players. Higher public awareness.	Unlikely or more contested. Disengagement likely characterised by limited roll back of policy and mirroring of EU policy even though no longer a member.

Policy

When the UK joined the EU, UK environmental policy making was ad hoc, reactive, and based upon end-of-pipe solutions (Weale *et al.* 2000, p. 177). Implementation was patchy, favouring a voluntarist approach that relied upon negotiated consent (Lowe and Ward 1998). This policy style was completely at odds with that favoured by the 'green pioneers' (Denmark, the Netherlands and Germany). These states uploaded domestic models to the European level to minimise their downstream implementation costs (Börzel 2002). Thus, the UK persistently found itself having to implement policies designed for other political systems, most often the more inflexible and legalistic German policymaking system (Wurzel 2006). Consequently, EU membership prompted the development in the UK of a more organised, proactive and integrated environmental policy based upon clear and enforceable targets. However, Jordan (2004) suggests that the extent to which the UK environmental policy paradigm shifted is debateable: in some areas, such as water and air, the EU's impact was more extensive than in others, such as land use planning (Wurzel 2006; Cowell and Owens 2016). Hence, the patterns of Europeanisation across the sector varied.

Nevertheless, over time UK policymakers gradually learnt to 'think European' (Jordan 2003, p. 263), as the government and civil servants sought to adapt to the challenge of policymaking in Brussels. From the 1990s onwards, UK actors sought, in certain policy areas, to set the EU's agenda, both by leading and blocking action. For example, the UK blocked major Commission proposals on environmental taxation and soil protection. In the early 1990s, the UK tried to shape the EU policy agenda in a range of ways. It sought to advance its voluntary policy approach, but met with limited success. UK attempts to shape the types of instruments used by the EU, by advancing the use of more market-based approaches such as environmental auditing and eco-labels, and integrated pollution control were better received (Jordan 2002; Haigh 2015). The UK also started to push for greater attention to be paid to the regulatory burdens (or 'red tape') of environmental action. This agenda became central to UK European policy, from Labour, which put 'better regulation' at the heart of its 2005 Presidency of the Council of the EU (House of Lords 2005), to the Conservatives. 'Cutting EU red tape' was also a core part of the 2010–2015 coalition government's EU strategy (Business Taskforce 2013) and it became one of David Cameron's four negotiation objectives when he set out to reform the EU Treaties prior to the 2016 Brexit referendum (Cameron 2015).

Politics

EU membership has shaped the strategies of various domestic actors (including non-government organisations (NGOs), businesses and to a lesser extent political parties). ENGOs have learned how to use EU governance structures to hold the UK government to account (Haigh 2015). For example, they contested UK attempts to minimise compliance costs of the Bathing Water Directive by designating only 27 bathing areas (Jordan 1997). The EU also provided funding for ENGOs and provided a platform for pan-European collaboration (Berny 2016). UK-based ENGOs have been central to the creation of EU-level ENGOs, which have influenced European policymaking (Berny 2008). Moreover, UK ENGOs and think-tanks have directly shaped EU policymaking, notably on integrated pollution control (Haigh 2015) and agricultural reform (Fouilleux and Ansaloni 2016).

UK businesses have also engaged with the EU's environment agenda. The desire to reduce costs and uncertainty for businesses has been a key driver of EU environmental policy. The Stern Review (Stern 2006), which made the case for ambitious climate policy, attracted wide business support in favour of EU climate regulations (Carter and Jacobs 2014). However, some companies and types of business continue to rail against Brussels 'red tape'. Developers have consistently identified the habitats and birds directives as imposing costs upon them. Farmers have levelled similar criticisms at EU pesticide regulations (National Farmers' Union 2017).

For the first 30 years of the UK's EU membership, the two main parties did not directly compete over the environment (Carter 2006). However, in the mid-2000s, the Conservative party sought to rebrand itself by embracing the environment as a way to 'detoxify' its image (Carter 2009). Labour, the Liberal Democrats and the Conservatives competed fiercely over climate change policies, leading to the adoption of the 2008 Climate Change Act (Carter and Jacobs 2014), which eventually proved divisive within the Conservative Party. The government's efforts to meet its EU renewables targets prompted internal opposition, particularly to onshore wind farms and green duties on energy bills (Carter and Clements 2015). Similarly, Conservative Chancellor George Osborne, identified the EU Habitats Directive as 'placing ridiculous costs on businesses' (Osborne 2011). Growing climate scepticism and opposition on the right to environmental regulations connected directly with and drew much strength from the powerful anti-EU lobby within the party.

Of UK political parties, the Greens have most obviously benefitted from EU membership. Since the introduction of proportional representation for European elections, the Greens have consistently secured representation of at least two Members of the European Parliament (MEPs). They have used their MEPs to build their domestic profile and secure further support (Bomberg and Carter 2006). For example, party co-chair Caroline Lucas used her stint as an MEP as a springboard to become a Westminster MP.

Polity

Whilst the EU has profoundly influenced the type of environmental policies implemented in the UK, the use of directives, 'orientated towards the ends to be achieved, rather than the means of achievement' (Bulmer and Jordan 2016, p. 9), has limited the EU's impact on domestic government. Rather, EU membership triggered a long and slow transformation within UK government to enable it to deal with its European neighbours (Burch and Bulmer 2005). Hence, whilst there has been reorganisation of government structures within the UK, particularly for environmental policy,[2] these changes have largely been driven by domestic concerns and it is challenging to disentangle the effect of Europeanisation from other processes (Bulmer and Jordan 2016; Jordan 2002). The most significant changes have occurred at the intersection of polity and policy where the act of pooling sovereignty has led to a pooling of capacity across EU states. Thus, several EU agencies are responsible for giving advice and administering EU rules, which means that EU states do not have to maintain equivalent structures at the domestic level.

The EU has also provided governance architecture to support policy development and implementation, including networks to exchange best practice such as the EU Network for the Implementation and Enforcement of Environmental Law (IMPEL). For example, the UK government is obliged to report on a whole range of policy activities, with the data made publicly available, as in the case of bathing waters. Furthermore, if the government fails to report regularly or to implement EU policies it can be held to account in national courts and see itself pursued at EU level. EU membership has consequently had a profound impact both upon the substantive norms underpinning environmental policy (what the rules are) and upon institutional norms (who enacts policy) (Roger 2016).

Furthermore, whilst EU membership has not profoundly affected government structures, the UK's devolution settlements have led to changes that EU membership has augmented. Currently, the EU sets minimum levels for environmental policy. Member states can diverge from them (under Article 193 of the Treaty on the Functioning of the EU—the so-called environmental guarantee), but only by pursuing higher environmental standards. The combined effects of Article 193 and the devolution settlements have allowed policy divergence to emerge within the UK where policies go further than the EU prescribed minimum (Hunt et al. 2016; Reid 2016). Consequently, Wales and Scotland have adopted more ambitious climate change policies than England (Royles and McEwen 2015); Wales, Scotland and Northern Ireland, have banned the cultivation of GMOs, but England has not (Coghlan 2015).

The referendum campaign and its aftermath: de-Europeanisation?

Overall we can see that policy, politics and polity have all been Europeanised, although the patterns and depth of that Europeanisation has varied. Below we review evidence from the referendum campaign and its aftermath (using our three categories) to determine the extent to which a demand for de-Europeanisation has existed, and, if so, the factors shaping it.

De-Europeanised policy?

A vocal lobby has emerged in the UK that favours rolling back EU legislation; Brexit has been presented as a golden opportunity to remove legislation that this group views as problematic, most notably the habitats and birds directives, which have become the *bête verte* of the conservative right (Environment Analyst 2016). Indeed, a key theme emerging from our stakeholder events was a deep fear that these two directives will face amendment or be removed after Brexit. Wildlife NGOs view UK legislation as offering weaker protections than its EU equivalent.[3] Michael Gove, appointed Secretary of State for the Environment, Food and Rural Affairs in Summer 2017, committed himself to the pursuit of a green Brexit, stating that there would be no weakening of standards post-Brexit (Gove 2017). However, the Department for International Trade will lead negotiations to secure post-Brexit trade deals. A clear fear that stakeholders expressed was that future trade deals would see products coming into the UK subjected to lower standards, leading to downward pressure upon domestic standards. Moreover, several interviewees, whilst welcoming a committed environment minister, were concerned that Mr Gove would not remain long in post, potentially leaving the environment vulnerable to the appointment of a less environmentally committed minister (Interviews 11/08/17; 07/09/17; 25/09/17). These comments flag a key concern that the previously stable policy regime emanating from Brussels could, when repatriated, be subject to domestic political variation (Interview 07/09/17). Stakeholders also expressed concern that the UK faces a reassertion of the voluntarist, flexible policy style preferred prior to the UK's EU membership, and still apparent in some areas such as planning (Cowell and Owens 2016).

To pave the way to Brexit the UK parliament has adopted an EU Withdrawal Act (EUWA), which has also led to concerns about environmental policy dismantling. When the UK joined the EU, it adopted the 1972 European Communities Act (ECA) to give effect to its obligations as an EU member state. In the UK, national implementing measures give directives legal effect; however, article 2(2) of the ECA gives regulations effect without any specific implementing legislation. Consequently, when the ECA ceases to apply, national measures adopted to give effect to directives would remain in place but regulations would cease to apply, potentially creating numerous regulatory gaps. To address this problem, the EUWA retains all existing EU rules and regulations in national law to prevent legal uncertainty in the short term.

This approach contains an assumption that at some point the vast array of laws adopted to give effect to EU obligations will be reviewed, and decisions made about whether they should be maintained or scrapped, raising the risk of policy dismantling. The Department for Environment, Food and Rural Affairs (Defra) is the second most heavily affected department by the EUWA, with the EU affecting approximately 80% of its work (National Audit Office 2017). There is an expectation that Defra will need to bring forward 95 statutory instruments to implement the EUWA, and two new domestic bills on agriculture and fisheries to replace the Common Agriculture and Common Fisheries policies (National Audit Office 2017). The scale of work Defra and other departments face is likely to mitigate against an immediate and widespread roll back of EU environmental policy. Equally, however, there is a risk that the Withdrawal Act has failed to sufficiently address regulatory gaps, or will leave unanticipated gaps within UK environmental governance. The fact that the EUWA, as originally framed, failed to copy over environmental principles from the EU treaties into UK law, such as the protection, precautionary and polluter pays principles, was identified by ENGOs as problematic. They campaigned vigorously on the issue leading the government to bring forward a consultation on environmental governance and principles (DEFRA 2018a) and the House of Commons adopted an amendment to the EUWA requiring the government to bring forward a Bill enshrining environmental principles (House of Commons 2018). However, the amendment to the EUWA does not demand the pursuit of a higher level of environmental protection, as required by the EU Treaties (see House of Commons 2018).

Moreover, the EUWA affords a good deal of discretion to government ministers to change laws without proper parliamentary scrutiny. One interviewee expressed concern that ministers would not be able to avoid the temptation to change laws and could do so without scrutiny and in ways that might appear minor but could be significant in scope, by, for example, removing all future reporting requirements from EU legislation being copied over (Interview 7/09/17).

Finally, it is worth noting that Brexit has implications for EU-level policy dynamics. The UK has been a strong advocate of higher ambition in areas such as climate change in the EU Council; Brexit may weaken the EU's ability to promote higher international standards (Interviews 07/09/17; 25/09/17). Conversely, the UK has also led calls to cut EU regulatory burdens; Brexit will test whether better regulation is merely an 'Anglo-Saxon obsession' (House of Lords 2005, p. 21). Brexit also constitutes one more policy challenge that the EU must face in its crowded policy agenda, raising the prospect that the environment will be pushed further down the list of policy priorities (Burns and Tobin 2016; Slominski 2016).

De-Europeanised politics?

ENGOs responded rapidly and robustly to Brexit by launching a Greener UK campaign and mobilising to call for a New Nature Protection Act to preserve and develop environmental protections (Greener UK 2017). When appointed, Michael Gove moved quickly to secure ENGO support by meeting with key personnel (Interview 31/08/17). However, it remains to be seen whether ENGOs will be able to maintain a high level of issue salience across such a wide range of issues. In the past, British ENGOs have also played a substantial role in supporting ENGOs in other EU member states (Hofmann 2019). Although this role may not completely disappear after Brexit, UK ENGOs may well be tempted to focus their energies on holding the line in the UK. Here the level of public awareness and the ability of ENGOs to mobilise the public will be central. Evidence suggests that the British continue to care about the environment. Prior to the referendum, 47% of the British public claimed to support more EU integration in environment and climate policy (Vasilopoulou 2015), and 46% supported a call for UK environmental protections to be stronger post-Brexit (YouGov 2016).

Of the main political parties, key Labour voices have emerged on the environment, most notably Sadiq Khan, the Mayor of London. Khan (2017) has sought to exploit London's poor air quality as a political issue. Labour garnered support from younger voters (60% of 18–24-year olds voted Labour) in the 2017 election (Holder *et al.* 2017); one strategy it may employ to win power is to use the environment agenda to create space between the Conservative and Labour parties. Commentators have suggested that the need to attract younger voters partly explains why the Conservative government published its long awaited 25-year environment plan (25YEP) in January 2018 (Steffani and Cooper 2018). Hence, whilst the environment did not feature much in the referendum, Brexit has seen a growing politicisation over environmental issues.

Within industry, the picture is even less clear. The Confederation of British Industry (CBI) has generally supported EU membership and the EU's green growth agenda. Concerns exist about the implications of Brexit for different UK economic sectors—leading to calls for a 'whole economy' approach (CBI 2016). Business has also expressed concern about uncertainty, especially over energy and product regulations (House of Commons 2017a). Sectors, such as chemical production, that are heavily reliant on EU rules to trade with the Single Market have been especially vocal about ensuring regulatory certainty. The UK government has indicated its desire to establish its own set of procedures and a new, national chemicals agency (DEFRA 2018b). However, MPs have emphasised that, to ensure frictionless trade into the Single Market, the pragmatic option is to abide by current EU product standards (even though the UK would have no role in determining them) and to accept EU regulatory oversight (House of Commons 2017b).

De-Europeanised polity?

The chemicals example highlights a much bigger issue: to what extent will actors within the polity still be able to draw upon the knowledge that has been pooled with EU partners? This question goes well beyond the chemicals sector to include a much larger number of trade substances (e.g. genetically modified organisms, foodstuffs, plastic waste). The expected loss of access to expertise combined with the civil service's diminished size, despite recent appointments, raises the risk that, whilst immediate wide-scale policy dismantling may not occur, significant national policy innovation is also unlikely.

Ironically, given the EU's limited direct impact upon how the UK government operates, Brexit is raising significant constitutional challenges to the existing relationship between different government tiers within the UK. The UK Government announced that the Withdrawal Act would provide delegated statutory powers to 'enable Ministers to adjust the *acquis* to fit the outcome of the negotiation' (Caird 2016, p. 24). This highlights the issue of veto players and the balance of power between the executive and legislative branches, and between Westminster and the devolved nations. It is unclear who will be responsible for sorting through the hastily carried over swathe of EU legislation. It remains to be seen whether ministers will be able to dismantle environmental policy through administrative channels, without parliamentary scrutiny (House of Lords 2017). There is also ongoing uncertainty about how UK-wide agreements on future environmental policy will be designed. The wording of the Withdrawal Act caused controversy between the UK government and the devolved administrations as the UK government suggested it would decide which policies would be devolved post-Brexit, which the Welsh and Scottish governments characterised as a power grab. In 2018 the Scottish government refused its consent to the EUWA and started preparing rival ('continuity') legislation (Scottish Government 2018). A major constitutional crisis is brewing between the UK polities over who has the right to decide on common UK environmental frameworks post-Brexit (Petetin 2018).

These polity-related issues may interact with politics as the veto players in the devolved nations form alliances with other actors to contest de-Europeanisation of environmental policy. One possible outcome is increasingly divergent policy patterns across the UK regions (once common EU rules are removed) (Reid 2015) and thus more varied patterns of (de) Europeanisation. A key limitation that the devolved nations face in pursuing different policy goals from the UK is their relatively weak capacity and expertise. Thus the UK faces the challenge of disentangling itself from the EU *acquis* and developing new patterns to share competence and develop policy across the UK nations, all under a severely constrained administrative capacity at every governance level. The government announced cuts to Defra's budget of 15% in real terms between 2015 and 2020; the Royal Society for the Protection of Birds and the Wildlife Trusts argue that Defra has borne some of the biggest cuts across the whole of Whitehall (Howard 2015). Whilst the government has since spent £2 billion in getting Whitehall ready for Brexit, extra funding for Defra has only partially offset previous and planned cuts, with staffing back to its 2011 level (Owen *et al.* 2018). Local government, which is often tasked with implementing environmental policy on the ground, has faced an equivalent level of budgetary restriction, with no extra Brexit preparation funding; these actors have hotly contested the availability of council policy instruments and budgetary resources, for instance to implement the 2017 national plan to improve town and city air quality (Merrick 2017).

Concerning governance capacity more broadly, a key risk of the EUWA is that laws pasted into the UK statute book may remain in place with no governance or legal infrastructure to support them. For example, the Water Framework Directive has reporting requirements obliging states to send regular updates on how they implement the directive. Such obligations may no longer apply to the UK; currently no detailed plans exist to replace these reporting requirements at the domestic level. Additionally, actors can currently pursue the government through the Courts for failing to implement EU legislation and ultimately in front of the Court of Justice of the European Union (CJEU). Thus, Client Earth successfully took the UK government to court for failing to implement air quality laws (Client Earth 2017). Once the UK leaves the EU that ability to rely upon the accountability and legal infrastructure enshrined by the EU to enforce legislation will no longer apply (House of Lords 2016). The government's 25YEP committed to bringing forward a new environmental watchdog for England to address these concerns, but its scope and power, and relationship to equivalent bodies in the devolved nations remain uncertain (DEFRA 2018b).

De-Europeanisation after Brexit?

Do these post-referendum discussions imply that the UK is moving towards a period of de-Europeanisation? Two broader points are worth making upfront. First, disentangling the UK from the EU will be challenging; if there is de-Europeanisation, it is unlikely to be rapid. One obvious exception would be if the UK fails to secure a deal or a suitable transition period, which could have significant effects in those policy areas where the UK relies upon EU expertise, and is unlikely to be able to put in place equivalent structures. A key example is the nuclear energy sector where, if the UK fails to reach an agreement with the EU, the UK may find itself struggling to source medical isotopes and nuclear fuel, notwithstanding the fact that the House of Commons has adopted a nuclear safeguards bill (Institute for Government 2018). Second, the patterns of de-Europeanisation are likely to be as 'differentiated' as the patterns of Europeanisation (Jordan and Liefferink 2004), which means we need to pay careful attention to changes within and the interactions between the policy, politics and polity dimensions.

The impact of Brexit and de-Europeanisation are consequently likely to vary: unpicking the policy patchwork of EU membership (Héritier 1996) could be complex and uneven. Copeland (2016) notes that the fact that employment policy was centralised facilitated de-Europeanisation. Environmental policy, with its messier, devolved structure will prove more resistant. In policy areas where Europeanisation is most profound, divergence may only emerge over the longer term. Given the EU's ability to impose its product standards on non-member states (i.e. all chemicals imported into the EU must be compliant with the regulation on Registration, Evaluation and Authorisation of Chemicals [REACH]), de-Europeanisation will be challenging. At the paradigmatic level, given the relative resilience of the UK's preferred policy approach (Jordan 2002), de-Europeanisation in the form of a reassertion of the UK style of policy at the domestic level is likely

Within the *politics* sphere, Brexit will shape various organisations, perhaps forcing a more introspective focus. Businesses have started mobilising within the UK, and expressed concern especially about the impact on investment and productivity (Cox *et al.* 2017, Savage 2017). There is scope for ongoing politicisation of environmental politics: what has been a technical matter decided in Brussels may become part of domestic political debate and contestation. Clear space between the parties emerged on the environment in the 2017 General Election (Laville *et al.* 2017). An on-going attempt to stigmatise 'EU green tape' occurred in the run up to the referendum campaign (Interview 31/08/17). If this campaigning continues, greater de-Europeanisation may occur. Here we depart from Copeland who assumes that wider public engagement and contestation act as a brake on de-Europeanisation; it may also be a driver.

Finally, on *polity*, the EU has had limited effects but the interaction between devolution and the EU's environmental guarantee has opened up the possibility of a differentiated de-Europeanisation and divergence across the composite states of the UK.

Overall, the future patterns of environmental governance look complex; ironically, the interaction between polity (where Europeanisation was most limited) and policy may make de-Europeanisation harder to achieve. Moreover, UK-level capacity (or rather its absence) is likely to be a central condition determining whether we see de-Europeanisation or disengagement in the environmental policy sector. Whilst the act of leaving the EU will be deliberate, the future patterns of policy change may end up being the result of disengagement rather than deliberate de-Europeanisation. Future studies may wish to distinguish between disengagement occurring where de-Europeanisation has been tried and has failed, and disengagement due to lack of capacity, energy, or political will. We suggest extending the de-Europeanisation typology to encompass failed de-Europeanisation, as distinct from passive de-Europeanisation (or *qua* Copeland disengagement).

Returning to Table 1, our initial expectations are being borne out—albeit with some additional consideration of the role of capacity. A limited capacity to roll back policy *or* to innovate is likely in the UK: we certainly see little appetite for or prospect of an environmental policy renaissance in the immediate future, with all that implies for environmental outcomes (Simkins 2017). Hence a discussion of capacity is essential when discussing the scope for de-Europeanisation. Building upon Table 1 and our analysis, we suggest a typology of disengagement that distinguishes between passive and failed de-Europeanisation and identifies capacity as a central variable (Table 2).

Conclusions

Brexit is the latest in a series of crises to beset the EU at a time when environmental policy was already struggling and in some respects, in retreat. As the UK is one of the EU's most Euro-sceptic states, it would be easy to assume that Brexit will herald wide-scale environmental policy dismantling at the national level. We contend, however, that, if and when the UK leaves the EU, this outcome is unlikely. UK policy has been profoundly Europeanised in ways that will be difficult to disentangle and completely reverse.

Table 2. Brexit as disengagement.

Expectation	Policy	Politics	Polity
Disengagement as passive de-Europeanisation.	Policy stays in place and gradually becomes outdated.	De-politicization and identification of environment as technical low politics. Lack of resources amongst NGOs.	Lack of state capacity to review or retrench policy.
Disengagement as failed de-Europeanisation	Deliberate attempts to de-Europeanise thwarted—policy stays as is, gradually being undermined via technical adjustments.	Contestation results in stalemate. Brexit fatigue sets in.	Joint-decision trap—too many veto players to move either way. Policy stasis follows.

Our analysis of debates about the future of environmental policy, interviews and stakeholder engagement leads us to expect capacity (or rather its absence) to be a central variable. Lack of capacity means that there will be limited ability to unpick the domestic manifestation of the EU's environmental *acquis*. Yet significant national policy innovation is also unlikely. This finding is consistent with earlier Europeanisation studies, which identified capacity as a limit upon policy implementation (Börzel 2002). Brexit may, therefore, have a relatively limited impact on policy outputs, but have potentially more significant consequences for policy outcomes, as stasis and passive environmental policy dismantling emerge over the medium to long term. The politics of the environment will be central: the UK's vibrant ENGO sector and deep-seated public support for protecting the environment will be crucial in determining the strength of post-Brexit environmental governance arrangements.

What are the implications for the future of EU environmental policy? In the absence of the UK, we may see less EU-level emphasis upon 'regulatory burdens' and the red tape agenda. We may also see other Euro-sceptic and environmentally-sceptic voices emerging to replace the UK, crucially in the area of climate change where the UK generally played an important role. However, our analysis suggests that the EU's impact upon its member states is such that disentangling and unpicking effects of membership is challenging, especially in a devolved polity. Hence, rather than heralding the disintegration of the EU and large-scale de-Europeanisation, Brexit may actually demonstrate the resilience of Europeanisation in face of extraordinary challenges. Returning to Hogwood and Peters (1982), we suggest that, whilst the UK is leaving the EU, there will be extensive and sticky policy residue that even the most committed Brexiteers will find challenging to remove.

Acknowledgements

We thank three anonymous referees, Sina Leipold and Andrea Lenschow for their detailed criticisms. We also thank participants in the Pisa ECPR workshop, 24–29 April 2016, 'The Future of Environmental Policy in the European Union Workshop', University of Gothenburg, 19–20 January 2017, and the panel at the 3rd International Conference on Public Policy (ICPP), Singapore, 28–30 June 2017. We are grateful to the ESRC for funding the research underpinning this work through its 'UK in a Changing Europe Initiative' (a Commissioning Grant, and Brexit Priority Grant ES/R00028X/1).

Notes

1. See https://www.cieem.net/data/files/Resource_Library/Policy/Policy_work/CIEEM_EU_Directive_Summaries.pdf for detail.
2. In 1997 the UK Labour government created the Department of Environment Transport and the Regions, which the Department for Environment Food and

Rural Affairs (Defra) replaced in 2001. It established the Department for Energy and Climate Change in 2008, which was dismantled and its functions redistributed by the Conservative government in 2016.
3. Wilkinson, Friends of the Earth Referendum Roadshow, Leeds, 13 May 2016.

Disclosure statement

No potential conflict of interest was reported by the authors.

Funding

This work was supported by the ESRC [ES/R00028X/1].

ORCID

Charlotte Burns http://orcid.org/0000-0001-9944-0417
Viviane Gravey http://orcid.org/0000-0002-3846-325X
Andrew Jordan http://orcid.org/0000-0001-7678-1024
Anthony Zito http://orcid.org/0000-0002-2312-4781

References

Aydın-Düzgit, S. and Kaliber, A., 2016. Encounters with Europe in an era of domestic and international turmoil: is Turkey a de-Europeanising candidate country? *South European Society and Politics*, 21 (1), 1–14. doi:10.1080/13608746.2016.1155282

Bauer, M.W., et al., 2012. Dismantling public policy: preferences, strategies, and effects. In: M.W. Bauer, C. Green-Pedersen, A. Héritier, and A. Jordan, eds. *Dismantling public policy*. Oxford: Oxford University Press, 203–226.

Berny, N., 2008. Le lobbying des ONG internationales d'environnement à Bruxelles. *Revue francaise de science politique*, 58 (1), 97–121. doi:10.3917/rfsp.581.0097

Berny, N., 2016. Environmental groups. In: C. Burns et al. The EU Referendum and the UK environment: an expert review, 112–124. Available from: http://environmenteuref.blogspot.co.uk/p/the-report.html [Accessed 1 January 2017].

Bomberg, E. and Carter, N., 2006. The greens in Brussels: shaping or shaped? *European Journal of Political Research*, 45 (S1), 99–125. doi:10.1111/j.1475-6765.2006.00651.x

Börzel, T.A., 2002. Pace-setting, foot-dragging, and fence-sitting: member state responses to Europeanization. *Journal of Common Market Studies*, 40 (2), 193–214. doi:10.1111/1468-5965.00351

Börzel, T.A. and Risse, T., 2003. Conceptualizing the domestic impact of Europe. In: K. Featherstone and C. Radaelli, eds. *The politics of Europeanization*. Oxford: Oxford University Press, 57–80.

Bulmer, S. and Jordan, A., 2016. National government. In: C. Burns et al. The EU Referendum and the UK environment: an expert review, 79–89. Available from: http://environmenteuref.blogspot.co.uk/ [Accessed 1 January 2017].

Burch, M. and Bulmer, S., 2005. The Europeanization of UK government: from quiet revolution to explicit step-change? *Public Administration*, 83 (4), 861–890. doi:10.1111/j.0033-3298.2005.00481.x

Burns, C. and Tobin, P., 2016. The impact of the economic crisis on European Union environmental policy. *Journal of Common Market Studies*, 54 (6), 1485–1494. doi:10.1111/jcms.12396

Business Taskforce, 2013. *Cut EU red tape - Report from the business taskforce*, 1–60. https://www.gov.uk/government/publications/cut-eu-red-tape-report-from-the-business-taskforce [Accessed 26 2 2018].

Caird, J., 2016. Legislating for Brexit : the great repeal bill. *Briefing Paper House of Commons Library*, 7793, 57.

Cameron, D., 2015. *A new settlement for the United Kingdom in a reformed European Union. Communication to Donald Tusk*. Available from: https://www.gov.uk/government/uploads/system/uploads/attachment_data/file/475679/Donald_Tusk_letter.pdf [Accessed 27 July 2017].

Carter, N., 2006. Party politicization of the environment in Britain. *Party Politics*, 12 (6), 747–767. doi:10.1177/1354068806068599

Carter, N., 2009. Vote blue, go green? Cameron's conservatives and the environment. *The Political Quarterly*, 80 (2), 233–242. doi:10.1111/poqu.2009.80.issue-2

Carter, N. and Clements, B., 2015. From 'Greenest government ever' to 'Get rid of all the green crap': David Cameron, the conservatives and the environment. *British Politics*, 10 (2), 204–225. doi:10.1057/bp.2015.16

Carter, N. and Jacobs, M., 2014. Explaining radical policy change: the case of climate change and energy policy under the British labour government 2006–2010. *Public Administration*, 92 (1), 125–141. doi:10.1111/padm.12046

CBI, 2016. *Making a success of Brexit: a whole-economy view of the UK-EU negotiations*. Available from: http://www.cbi.org.uk/insight-and-analysis/making-a-success-of-brexit/ [Accessed 18 June 2017].

Client Earth, 2017. *High court judgment on air pollution a 'shot across the bows' of government*. Available from: https://www.clientearth.org/high-court-judgment-air-pollution-shot-across-bows-government/ [Accessed 19 July 2017].

Coghlan, A., 2015. More than half of EU officially bans genetically modified crops. *New Scientist*, 5 October 2017.

Copeland, P., 2016. Europeanization and de-Europeanization in UK employment policy: changing governments and shifting agendas. *Public Administration*, 94 (4), 1124–1139. doi:10.1111/padm.2016.94.issue-4

Cowell, R. and Owens, S., 2016. Land use planning. In: C. Burns et al. The EU Referendum and the UK environment: an expert review, 57–67. Available from: http://environmenteuref.blogspot.co.uk/. [Accessed 1 January 2017].

Cox, J., Chu, B., and Rodinova, Z., 2017. Cost of Brexit: the impact on business and the economy so far. *The Independent*. 21 April.

DEFRA, May 2018a. *Environmental principles and governance after the United Kingdom leaves the European Union, consultation on environmental principles and accountability for the environment*. Available from: http://www.gov.uk/government/publications [Accessed 22 June 2018].

DEFRA, January 2018b. *A green future: our 25 year plan to improve the environment*. Available from: https://www.gov.uk/government/publications/25-year-environment-plan [Accessed 11 January 2018].

Environment Analyst, 2016. *Fears grow for nature directives post-Brexit*. Available from: https://environment-analyst.com/48344/fears-grow-for-nature-directives-in-the-uk-post-brexit [Accessed 18 June 2017].

Falkner, G., 2016. The EU's current crisis and its policy effects: research design and comparative findings. *Journal of European Integration*, 38 (3), 219–235. doi:10.1080/07036337.2016.1140154

Fouilleux, E. and Ansaloni, M., 2016. The common agricultural policy. *In*: M. Cini and N. Pérez-Solórzano Borragán, eds. *European Union politics*. Oxford: Oxford University Press, 308–322.

Gove, M., 2017. *The unfrozen moment – delivering a green Brexit*, Speech. Available from: https://www.gov.uk/government/speeches/the-unfrozen-moment-delivering-a-green-brexit [Accessed 1 Janury 2018].

Gravey, V. and Jordan, A., 2016. Does the European Union have a reverse gear? Policy dismantling in a hyperconsensual polity. *Journal of European Public Policy*, 23 (8), 1180–1198. doi:10.1080/13501763.2016.1186208

Greener UK, 2017. Greener UK. Available from: http://greeneruk.org/ [Accessed 18 June 2017].

Haigh, N., 2015. *EU environmental policy: its journey to centre stage*. London: Routledge.

Héritier, A., 1996. The accommodation of diversity in European policy-making and its outcomes: regulatory policy as a patchwork. *Journal of European Public Policy*, 3 (2), 149–167. doi:10.1080/13501769608407026

Hofmann, A., 2019. Left to interest groups? On the prospects for enforcing environmental law in the European Union. *Environmental Politics*, 28 (2).

Hogwood, B. and Peters, B.G., 1982. The dynamics of policy change: policy succession. *Policy Sciences*, 14 (3), 225–245. doi:10.1007/BF00136398

Holder, J., Barr, C., and Kommenda, N., 2017. Young voters, class and turnout: how Britain voted in 2017. *The Guardian*, 20 June. Available from: https://www.theguardian.com/politics/datablog/ng-interactive/2017/jun/20/young-voters-class-and-turnout-how-britain-voted-in-2017 [Accessed 1 January 2018].

House of Commons, 2017a. *The future of the natural environment after the EU Referendum inquiry*, Environmental Audit Committee. Available from: https://publications.parliament.uk/pa/cm201617/cmselect/cmenvaud/599/599.pdf [Accessed 27 July 2017].

House of Commons, 2017b. *The future of chemicals regulation after the EU Referendum inquiry*, Environmental Audit Committee. Available from: https://www.publications.parliament.uk/pa/cm201617/cmselect/cmenvaud/912/912.pdf [Accessed 18 June 2017].

House of Commons, 2018. *European Union withdrawal bill, commons amendments in lieu, amendments to amendments and reasons*. Available from https://publications.parliament.uk/pa/bills/lbill/2017-2019/0111/18111.pdf#page=3 [Accessed 29 June 2018].

House of Lords, 2005. Ensuring effective regulation in the EU. *9th Report of Session 2005–2006*.

House of Lords, 2016. Brexit: environment and climate change inquiry. *European Union Committee. 12th Report of Session 2016–17, HL Paper 129*. Available from: https://www.parliament.uk/business/committees/committees-a-z/lords-select/eu-energy-environment-subcommittee/inquiries/parliament-2015/brexit-environment-and-climate-change/[Accessed 1 January 2017].

House of Lords, 2017. The 'great repeal bill' and delegated powers. *Constitution Committee, 9th Report of Session 2016–17. HL Paper 123*. Available from: https://

publications.parliament.uk/pa/ld201617/ldselect/ldconst/123/12302.htm [Accessed 30 October 2018].

Howard, E., 2015. Defra hit by largest budget cuts of any UK government department, analysis shows. *The Guardian*, 11 November. Available from: https://www.theguardian.com/environment/2015/nov/11/defra-hit-by-largest-budget-cuts-of-any-uk-government-department-analysis-shows [Accessed 27 July 2017].

Hunt, J., Minto, R., and Woolford, J., 2016. Winners and losers: the EU referendum vote and its consequences for Wales. *Journal of Contemporary European Research*, 12 (4), 824–834.

Institute for Government, 2018. *Explainers: euratom*. Available from https://www.instituteforgovernment.org.uk/explainers/euratom [Accessed 29 June 2018].

Jordan, A., 1997. 'Overcoming the divide' between comparative politics and international relations approaches to the EC: what role for 'post-decisional politics'? *West European Politics*, 20 (4), 43–70. doi:10.1080/01402389708425217

Jordan, A., 2002. *The Europeanisation of British environmental policy, a departmental perspective*. Basingstoke: Palgrave Macmillan.

Jordan, A., 2003. The Europeanisation of national government and policy: a departmental perspective. *British Journal of Political Science*, 33 (2), 261–282. doi:10.1017/S0007123403000115

Jordan, A., 2004. The United Kingdom. From policy 'taking' to policy 'shaping'. *In*: A. Jordan and D. Liefferink, eds. *Environmental policy in Europe: the Europeanisation of national environmental policy*. London: Routledge, 205–223.

Jordan, A., Bauer, M.W., and Green-Pedersen, C., 2013. Policy dismantling. *Journal of European Public Policy*, 20 (5), 795–805. doi:10.1080/13501763.2013.771092

Jordan, A. and Liefferink, D., 2004. The Europeanisation of national environmental policy. *In*: A. Jordan and D. Liefferink, eds. *Environmental policy in Europe: the Europeanisation of national environmental policy*. London: Routledge, 1–14.

Judge, D., ed., 1992. A green dimension for the European Community: political issues and processes. *Environmental Politics (Special Issue)*, 1, 4.

Khan, S. 2017. *Mayor of London Sadiq Khan's response to government air quality plan*. Available from: https://www.london.gov.uk/press-releases/mayoral/response-to-government-air-quality-plan [Accessed 1 January 2018].

Laville, S., Sauven, J., and Hogg, D. 2017. How do the four main parties compare on the environment? *The Guardian*, 21 May.

Lockwood, M., 2013. The political sustainability of climate policy: the case of the UK climate change act. *Global Environmental Change*, 23 (5), 1339–1348. doi:10.1016/j.gloenvcha.2013.07.001

Lorenzoni, I. and Benson, D., 2014. Radical institutional change in environmental governance: explaining the origins of the UK climate change act 2008 through discursive and streams perspectives. *Global Environmental Change*, 29, 10–21. doi:10.1016/j.gloenvcha.2014.07.011

Lowe, P. and Ward, S., eds., 1998. *British environmental policy and Europe: politics and policy in transition*. London: Routledge.

Merrick, R., 2017. Petrol-diesel car ban: government plan dismissed as 'smokescreen' after key air pollution policies dumped. *The Independent*, 26 July.

National Audit Office, December 2017. *Implementing the UK's exit from the European Union, the department for environment, food & rural affairs*. Available from: https://www.nao.org.uk/wp-content/uploads/2017/12/Implementing-the-UKs-exit-from-the-European-Union-the-Department-for-Environment-Food-Rural-Affairs.pdf [Accessed 1 January 2018].

National Farmers' Union, 2017. *NFU reaction to neonicotinoids announcement.* Available from: https://www.nfuonline.com/cross-sector/environment/bees-and-pollinators/bees-and-pollinators-news/nfu-reaction-to-neonicotinoids-announcement/ [Accessed 30 October 2018].

Osborne, G., 2011. *Autumn statement, 29 November 2011.* Hansard, Vol. 536, Column 799. Available from: https://hansard.parliament.uk/Commons/2011-11-29/debates/11112951000009/AutumnStatement?highlight=autumn%20statement#contribution-11112951000260. [Accessed 30 October 2018].

Owen, J., Llyod, L., and Rutter, J., 18 June 2018. *Preparing Brexit, how ready is Whitehall?* Institute for Government. Available from: https://www.instituteforgovernment.org.uk/sites/default/files/publications/IFGJ6279-Preparing-Brexit-Whitehall-Report-180607-FINAL-3c.pdf [Accessed 1 July 2018].

Petetin, L., 2018. International obligations and devolved powers – ploughing through competences and GM crops. *Environmental Law Review,* 20 (1), 16–31. doi:10.1177/1461452918759639

Raagmaa, G., Kalvet, T., and Kasesalu, R., 2014. Europeanisation and de-Europeanisation of Estonian regional policy. *European Planning Studies,* 22 (4), 775–795. doi:10.1080/09654313.2013.772754

Radaelli, C., 2004. Europeanisation: solution or problem? *European Integration On-Line Papers,* 8 (16). Available from: http://eiop.or.at/eiop/texte/2004-016a.htm [Accessed 1 January 2017].

Reid, C., 2015. *Written evidence to the environment audit committee's assessment of EU-UK environmental policy.* Available from: http://data.parliament.uk/writtenevidence/committeeevidence.svc/evidencedocument/environmental-audit-committee/assessment-of-euuk-environmental-policy/written/24036.pdf [Accessed 31 December 2016].

Reid, C., 2016. Brexit and the future of UK environmental law. *Journal of Energy and Natural Resources Law,* 34 (4), 407–415. doi:10.1080/02646811.2016.1218133

Roger, A., 2016. *Written evidence. Submitted to house of commons environmental audit committee inquiry on the future of chemicals regulation after the EU referendum.* Available from: http://data.parliament.uk/WrittenEvidence/CommitteeEvidence.svc/EvidenceDocument/Environmental%20Audit/EU%20Chemicals%20Regulation/written/45853.html [Accessed 1 June 2017].

Royles, E. and McEwen, N., 2015. Empowered for action? Capacities and constraints in sub-state government climate action in Scotland and Wales. *Environmental Politics,* 24 (6), 1034–1054. doi:10.1080/09644016.2015.1053726

Savage, M., 2017. Big business leaders press Theresa May to rethink hard Brexit. *The Observer,* 18 June.

Scottish Government., 2018. *Returning EU powers.* Available from: https://news.gov.scot/news/returning-eu-powers [Accessed 1 January 2018].

Simkins, G., 2017. Brexit 'could threaten' microbeads ban. *The ENDS Report.* Available from: http://www.endsreport.com/article/56595/brexit-could-threaten-microbeads-ban?printFriendly=true [Accessed 18 June 2017].

Slominski, P., 2016. Energy and climate policy: does the competitiveness narrative prevail in times of crisis? *Journal of European Integration,* 38 (3), 343–357. doi:10.1080/07036337.2016.1140759

Steffani, S. and Cooper, C., 2018. *The green washing of Theresa May.* Politico. Available from: https://www.politico.eu/article/the-greenwashing-of-theresa-may/. [Accessed 10 January 2018].

Steinebach, Y. and Knill, C., 2017. Still an entrepreneur? The changing role of the European Commission in EU environmental policy-making. *Journal of European Public Policy*, 24 (3), 429–446. doi:10.1080/13501763.2016.1149207

Stern, N., 2006. *The economics of climate change. The Stern review*. Cambridge: Cambridge University Press.

Vasilopoulou, S., 2015. Mixed feelings: Britain's conflicted attitudes to the EU before the referendum. *Policy Network Paper*. Available from: http://www.policy-network.net/publications_detail.aspx?ID=4964 [Accessed 18 June 2016].

Weale, A., et al., 2000. *European environmental governance*. Oxford: Oxford University Press.

Wurzel, R., 2006. *Environmental policy-making in Britain, Germany and the European Union*. Manchester: Manchester University Press.

YouGov. August 2016. *YouGov survey Brexit environment*. Available from: https://www.foe.co.uk/sites/default/files/downloads/yougov-survey-brexit-environment-august-2016-101683.pdf [Accessed 26 Febuary 2018].

Zito, A.R., Burns, C., and Lenschow, A., 2019. Is the trajectory of European Union environmental policy less certain? *Environmental Politics*, 28 (2).

Voluntary instruments for ambitious objectives? The experience of the EU Covenant of Mayors

Ekaterina Domorenok

ABSTRACT
Over the last decade, innovative governance architectures have been progressively promoted across European Union (EU) environmental and climate policies with the purpose of improving the effectiveness of intervention through better cross-sectoral policy integration and increased involvement of sub-state and non-governmental actors in the policy process. By combining the theoretical insights of polycentric governance and the concept of usage, the case of the Covenant of Mayors (CoM) is analysed to uncover the extent to which this voluntary programme has empowered local authorities within the EU strategy for sustainable energy by encouraging coordination and learning. This illustrates how a range of policy variables determined the dynamics of the programme's implementation in Italy and the United Kingdom.

Introduction

Previous research has illustrated how European environmental policies have changed over time by gradually enlarging the area of application of softer forms of regulation, along with the traditional and still dominant command-and-control style (Jordan and Adelle 2013). In particular, so-called New Environmental Policy Instruments (NEPIs) have contributed to integrating environmental concerns into sectoral policies, whilst simultaneously allowing the inclusion of new actors (in particular, enterprises and civil society) in environmental governance (Wurzel *et al.* 2013).

The European Union (EU) post-crisis policy agenda appears to have further strengthened this trend, as the Strategy Europe 2020 explicitly endorsed the need to increase cross-sectoral integration and the use of flexible policy instruments based on self-regulation and financial incentives (European Commission 2010). However, differently from NEPIs, which introduced specific organisational and procedural arrangements for implementing the principle of environmental policy integration across sectors, more recent policy programmes promote more strategic action within the framework of climate change mainstreaming (European Commission 2011). The logic of 'governing through enabling' clearly underpins these programmes (Kern and Bulkeley 2009), which target the local level in particular, aiming to improve complementarity between hard and soft policy tools and financial incentives with the objective of encouraging public authorities, private actors and civil society to take pro-active action in the EU strategy for combating climate change. Participation in collaborative networks, strong bottom-up commitment and motivation to obtain specific knowledge are supposed to be the main drivers of joint action.

Here, I analyse the Covenant of Mayors (CoM) programme, which the European Commission launched in 2008 to endorse and support local efforts in the implementation of sustainable energy policies. When joining the network, Covenant signatories commit to the EU Energy Package targets by pledging to prepare local Sustainable Energy Action Plans (SEAPs); these include measures to reduce carbon dioxide (CO_2) emissions and improve energy efficiency in a wide range of sectors that the Emissions Trading System (ETS) does not cover, such as public and residential buildings, transport, waste and lightening. Regular monitoring and benchmarking established by the CoM, and access to knowledge resources and capacity-building activities, aim to encourage coordination and enhance mutual trust among its participants.

The objective of creating a polycentric opportunity structure appears to inspire this design within the EU, which may act as a 'massive transfer platform' (Radaelli 2000, p. 6), for enhancing the voluntary climate action of local authorities for improving knowledge and practices in the field of sustainable energy. However, previous research has largely overlooked the expected benefits of the CoM; this scholarship mainly described CoM objectives and membership compared to other transnational municipal networks (TMN) operating within the multi-level system of European governance (Kern and Bulkeley 2009) or at global level (Hakelberg 2011, Bulkeley et al. 2012, Bouteligier 2013, Heidrich et al. 2016). Few studies have raised questions about CoM impacts on specific EU policy goals or the motivations of cities for getting involved (Busch 2015). Moreover, the twofold nature of the CoM has not been explored in depth: in contrast to other TMNs, the CoM combines a voluntary commitment by local authorities with a well-defined pattern of 'orchestration' (Abbott 2012) by EU institutions, including the coordination function of the European Commission (DG Climate), the scientific support of the Joint Research Centre (JRC), the multiple cross-cutting linkages between CoM activities and other EU policies and financial programmes, and a mechanism of exclusion of those who do not comply with the established obligations.

Against this background, a polycentric governance approach appears to offer a number of advantages for understanding the relevance of the CoM in the perspective of EU climate policy ambitions, although it needs further elaboration in order to capture how the CoM is enhancing local action across the different dimensions of the network.

Therefore, to be able to spell out 'opportunities and pitfalls' of the polycentric governance architecture (Jordan et al. 2015, 2018) that the CoM embodies, I suggest bridging the seemingly distant but actually complementary literatures on polycentric governance (Ostrom 2009, 2010) and the concept of usage (Jacquot and Woll 2003, Woll and Jacquot 2010). While coming from different theoretical and empirical backgrounds – the study of collective action in the former case and the research on social dynamics of European integration in the latter – these concepts provide mutually beneficial insights on how and why of policy actors become engaged in supranational systems of coordination.

I organise this contribution as follows. After a brief overview of the main theoretical assumptions and analytical premises, I analyse the CoM structure and functions and provide a comparative in-depth analysis of its implementation in Italy and the UK. I conclude by illustrating how we can improve our understanding of the effectiveness of polycentric governance architectures for climate change by analysing their actual usage by policy addressees.

Understanding EU climate change policies and governance: new perspectives on policy actors

Since the second half of the 1990s, the EU has been experimenting with modes of governance that are not based on law and hierarchy (Radaelli 2008, p. 239), with the objective of enhancing the alignment of national strategies with common objectives in those policy areas where the application of hard regulation has proven to be unfeasible or problematic. Notably, voluntary commitment, networking, coordination and learning were supposed to be the main drivers of the policy change (Radaelli 2008) within the strategy for sustainable development, which has significantly influenced the evolution of EU environmental policy (Jordan and Adelle 2013).

However, scholarship has underexplored the effectiveness of these instruments so far, as the dominant focus on institutional variables and the macro policy perspective (Radaelli et al. 2012) has limited our understanding of how target actors perform the process of policy implementation and why and to what extent they exploit opportunities offered by the EU political arena. In fact, the top-down approach that prevailed in research on policy implementation in the EU until recently (Treib 2014) assumes implementation problems were mainly problems of compliance, depending on the degree of institutional adjustment to EU pressures on national arrangements (Heidbreder 2017). Although studies on policy coordination (Jordan et al. 2005) and climate policy mainstreaming (Berkhout et al. 2015) have partially addressed these gaps, we need more research to understand how micro-policy foundations matter for the overall success of implementation of policies pursuing environmental sustainability as a cross-cutting priority. The polycentric design of the CoM hence provides an interesting example of experimentation in climate governance (Hildén 2017, Matschoss and Repo 2018).

The CoM programme aims to enable large-scale action for reducing CO_2 emissions and improve energy efficiency by involving different territorial levels in a system of mutual monitoring, incentives and guidance for policy innovation and strategic partnerships. More specifically, local authorities that voluntarily commit to the programme are required to formulate and implement their SEAPs in collaboration with relevant stakeholders. The programme also expects local authorities to carry out and make public monitoring of implemented actions, and encourage to develop and share benchmarks. Lastly, the JRC as well as individual CoM participants promote a series of joint training activities, with the objective of creating and transferring knowledge of sustainable energy policies.

Furthermore, this programme has embedded several learning enhancement mechanisms (Radaelli 2008, p. 245), including development of comparable statistics and common indicators for SEAPs, Basic Emission Inventories (BEIs) and monitoring, systematic diffusion of expertise and practical experience through benchmarking and training, and enhancing strategic use of policy linkages and development of a common policy discourse through joint events and internal communication. Additionally, by enabling open access to all relevant municipality documents on the programme website, the CoM aims to increase transparency and mutual trust among the participants. A system of territorial coordinators (Regions and Provinces) and supporters (networks, agencies, etc.) ensures a multi-level perspective. All these actors can take action in their territories and areas of expertise (e.g. energy, environment, water, air) involving different levels of governance (national, regional and local) in order to promote the CoM initiatives and support the commitments of its signatories.

As a consequence, successful implementation of the programme depends on the extent to which local authorities engage in its activities, comply with common policy guidance and commit to policy objectives and targets.

Previous studies on TMN for climate (Hakelberg 2011, Heidrich et al. 2016) have reviewed the composition and main characteristics of the CoM, and have also discussed its role as a form of transnational regulatory regime (Heyvaert 2013) or multi-level network of governance (Kern 2010). Nevertheless, scholars have made only a few attempts to investigate how cities are actually involved in the CoM activities and how membership correlates with the improvement of local climate strategies beyond the formal submission of SEAPs (Busch 2015). A systematic assessment of this programme as a tool of policy coordination and learning in the EU is still missing.

I address this gap by testing empirically the validity of theoretical assumptions about the potential of polycentric systems for governing climate change. This examination at the same time enables us to understand the impact of EU experimental forms of governance based on voluntary commitment and soft coordination.

Ostrom (1999) conceived polycentric systems as governance arrangements characterised by consistent and predictable patterns of interaction between multiple formally independent units operating at different scales; these units get involved in a process of mutual monitoring, learning and ultimately adaption of better strategies over time based on reciprocity and trust. Drawing on the seminal studies on polycentric climate governance (Ostrom 2009, 2010), I adopt a two-fold analytical perspective to understand whether and, under which conditions, the polycentric architecture embedded by the CoM can become a vector of policy change in the field of climate policies by providing local authorities with new policy knowledge, strategic opportunities and learning resources.

Hence, I assess the effectiveness of the CoM based on a range of variables that have been considered crucial for the success of polycentric systems (Ostrom 2009), including the following:

- Local authorities perceive the CoM objectives as important for their own achievements over the long term;
- Those involved consider monitoring and sanctioning to be feasible and appropriate;
- CoM members see it as a useful source of new knowledge and policy ideas;
- Participating municipalities perceive the CoM as a reliable source of information about the costs, benefits and improvements;
- Political commitment exists;
- Gaining a reputation for being a trustworthy reciprocator is important to those involved.

While offering useful insights on the overall potential of the CoM for enhancing local authorities' commitment to EU climate objectives, this framework does not allow us to trace and explain the dynamics of municipalities' performance across different activities of the programme.

We can address these issues with the help of action-theoretical considerations developed via the concept of 'usage of Europe' (Jacquot and Woll

2003, Woll and Jacquot 2010), which suggests that it is not an *a priori* 'degree of coercion' of policy instruments that matters, but the usage that is made of them, their concrete implementation and the meaning that actors attach to them by seizing the EU as a set of opportunities, be they institutional, ideological or organisational (Jacquot and Woll 2003). By distinguishing between three main types of usage (cognitive, strategic and legitimating), this concept highlights the relevance of interests and motivations guiding actors' behaviour in the context of EU policies, though lacking analytical linkages required for assessing the effectiveness of voluntary policy instruments and, in particular, those characterised by a polycentric architecture.

Therefore, building on the theoretical insights by Jacquot and Woll (2003), I suggest the following four scenarios of how local authorities may engage with the polycentric system of shared commitments and opportunities that the CoM established:

- *Symbolic* – occasional interest drives sporadic activation by signature for membership or submission of the SEAPs but no or limited involvement in monitoring, benchmarking.
- *Strategic* – authorities use the CoM for increasing resources and strategic opportunities (partnerships, funding); pragmatic motivations prevail to either upgrade local sustainable energy policies or increase local capacity of action and access to new policy resources.
- *Cognitive* – the objective is to obtain new ideas, acquire and/or improve knowledge and expertise related to new policy concepts in the field of sustainable energy; authorities intensively employ methodological and guidance resources.
- *Legitimating* – authorities use participation in the CoM to legitimise the pre-existing strategies; higher ambition of targets than those foreseen by the CoM; limited interest in knowledge resources, greater focus on visibility and political image.

Distinguishing between these scenarios can shed light on differences in behaviour of local authorities in domestic contexts, providing insights into strengths and weaknesses of the CoM as a tool for policy coordination and learning in EU climate policies.

Therefore, by comparing the implementation of the CoM in Italy and the United Kingdom, my objective is to understand whether there have been similarities in the implementation scenario of the CoM in these countries regardless of significant differences between these contexts in terms of at least two essential conditions determining local climate action: the advancement of national climate policies and the flexibility of local governance architectures. A number of advantages are apparently in place for more successful implementation in the UK, as this country has been an environmental leader within the EU (Wurzel et al. 2013), especially at the local level, and the presence of multiple public–public and public–private networks characterises its system of local governance. Italy, by contrast, has lagged behind in the implementation of EU environmental legislation (Jordan and Adelle 2013), and a hierarchical and inflexible pattern of territorial governance has limited its local authorities.

Research methodology

I operationalise the analytical framework by using a mixed quantitative-qualitative methods approach. I have collected and elaborated quantitative data from the CoM website for measuring variables from 1 to 5 on the CoM. More specifically, I have assessed the overall relevance of the CoM for improving sustainable energy polices (V1) as well as local authorities' political commitment to the initiative (V5) by considering the total number of municipalities that applied for participation, the timespan between the application and submission of the local SEAP (one year is considered to be an optimum time indicating a high commitment to the programme), the status of local SEAPs and the nature of their commitment. I have evaluated the propensity to consider mutual monitoring as feasible and important (V2) and the perception of the CoM as a reliable source of new policy knowledge (V3, 4) based on the share of SEAPs involved in monitoring and benchmarking activities as compared with the total number of submitted plans. These data cover the entire CoM membership in Italy and the UK, except for V3 for which I have selected a sample of 500 Italian municipalities, including municipalities of different size and geographic position.

In addition, I conducted a survey to investigate opinions and perceptions of local authorities on key aspects of the CoM for which quantitative data were not available, such as, for example, the relevance of economic and human resources for participation in the network, the possibility to acquire new knowledge, develop cooperation networks, increase international visibility and trust (V6). We sent a questionnaire to all participating municipalities in the UK and a sample of 100 Italian municipalities representing a great variety of local conditions in terms of size, geographic position and status in the CoM. A total of 35 respondents (28 from Italy and 7 from the UK) provided detailed feedback. Although the collected survey data are not comprehensive, they offer many important insights about the perceived strength and weakness of the CoM.

We have collected additional qualitative insights through 12 semi-structured interviews with sustainable energy and climate officers (see Appendix), which were carried out between 2015 and 2017 with the purpose of acquiring a detailed understanding of motivations of local authorities for being involved in the CoM. Interviewees were selected so as to cover the most different situations of municipalities in terms of geographic position, size, commitment status and degree of activeness in the network.

The Covenant of Mayors: a tool for climate policy mainstreaming in expansion

The EU Commission promoted the CoM to increase the visibility of the role of local authorities and enhance their contribution to the implementation of EU targets established by the Climate and Energy package. The idea that a polycentric pan-European policy network may increase knowledge of and commitment to sustainable energy strategies has underpinned the overall setting of the CoM, whilst favouring cooperation between different territorial authorities, and improving communities' and stakeholders' awareness of the issue of climate change at local level. The programme created several devices to guide change. As far as the production and exchange of knowledge are concerned, the Covenant of Mayors Office (COMO) of the European Commission and the JRC have jointly developed common templates and methodologies for local SEAPs, BEIs and monitoring, in order to provide a harmonised methodological framework. This also implied that specific expertise and capacities would be available or acquired at the local level. Furthermore, the CoM has established a coordinated mechanism for regular monitoring and reporting of progress in terms of greenhouse gas emissions (GHG) emissions, final energy consumption, clean energy production, estimated reductions and energy savings and required to publish the related local data on the CoM website with the purpose of increasing responsibility and awareness about the joint commitment among participants. Moreover, the EU Commission has stressed the necessity to strengthen territorial coordination, boost local public and private partnerships, introduce peer-to-peer mechanisms and foster capacity building among local authorities' staff and stakeholders (JRC 2015) by establishing a system of territorial coordinators and supporters, organising training activities and calling for the involvement of the private sector in the definition and implementation of SEAPs.

Thus, the CoM encourages local authorities to align their strategies to common objectives by benefitting from network resources and developing new expertise and practical knowledge about innovative sustainable energy policy solutions. The COMO holds the responsibility for implementation of the initiative, providing first-line technical and administrative assistance to signatories and facilitators.

The nature of the commitments for municipalities willing to participate in the CoM has evolved over time, becoming more and more ambitious.

Those who joined the network between 2008 and October 2015 committed through their SEAP to reach the target of 20% GHG reduction by 2020. Instead, the Mayor Adapt initiative launched in 2015 introduced a more ambitious target of 40% reduction by 2030 and the adaptation objective. Accordingly, starting from 2016, local authorities had to develop Sustainable Energy and Climate Action Plans (SECAPs), which contained adaptation and mitigation measures aiming to achieve the 40% emission reduction target by 2030. In 2016, the CoM signatories endorsed a shared vision until 2050: accelerating the decarbonisation of their territories, strengthening their capacity to adapt to unavoidable climate change impact and allowing their citizens to access secure, sustainable and affordable energy (CoM 2015).

The ambition of action and the geographical scope of the programme have increased over time, especially since adoption of the Energy Efficiency Directive (2012/27/EU), which specifically acknowledged the role of local governments and the CoM initiative in achieving significant energy savings (European Union 2012). The Directive called for Member States to encourage municipalities and other public bodies to adopt integrated and sustainable energy efficiency plans. The Directive stressed (preamble 18) the encouragement of the exchange of experience between cities, towns and other public bodies in order to develop more innovative experiences. In 2014, while highlighting the success of the CoM, the Commission confirmed its ongoing support for the initiative as an important platform for achieving progress, especially on energy efficiency in buildings. Furthermore, the European Commission's Energy Security Strategy called on Member States to accelerate the implementation of SEAPs in the 'stress test' countries as a means to improve the Union's security of supply. Finally, a number of EU financial instruments, such as ELENA, LIFE+ and Structural funds, supported participation in the CoM municipalities, (indirectly) providing grants for developing aspects of local sustainable energy strategies.

The membership of the CoM has increased: there were 7204 signatories, and the JRC approved 5679 SEAPs as of January 2017, with the highest number of participants from Italy and Spain. Recently, the geographical scope of the CoM has further extended, involving the Eastern Partnership, Central Asia and 10 southern Mediterranean countries. Regional offices in North and Latin America will soon follow, while at the beginning of 2017 the Commission launched a new partnership with the Compact of Mayors and the global Covenant of Mayors for Climate Change and Energy. This partnership strengthened the crosscutting links between the CoM and other transnational city networks, such as ICLEI, C40 or Climate Alliance, in which it has been strongly embedded since its very origins (Busch 2015).

The implementation of the CoM: comparative insights from Italy and the UK

Given the consolidation and expansion of the CoM, it is interesting to observe how differently local authorities perform within the network.

The Italian and UK contexts offer completely different conditions for the implementation of the CoM. Recently, the UK has been among the leaders in the implementation of environmental and climate change policies, with the UK Climate Change Act of 2007 imposing legally binding targets for the reduction of CO2 emissions by at least 50% by 2050. National regulations have strongly influenced local authorities, steering local climate action by specific assessment indicators (Kern 2010). Italy, by contrast, has one of the highest rates of EU environmental infringement procedures and its climate policies are embryonic at both national and local levels, with poor attempts at coordination (Lumicisi 2013).

Yet, the general interest in the CoM has been much higher in Italy with 3132 signatories as of August 2017 compared to 36 in the UK. As Figure 1 shows, there has been also a considerable divergence in terms of size and population of participating cities: in Italy small cities with a population of less than 10,000 inhabitants constitute the clear majority (75%) of its membership, covering around 65% of the country population. By contrast, in the UK, cities participating in the CoM are large, corresponding to 28% of total population.

The general commitment of municipalities to the CoM activities (Table 1) varies significantly between and within the countries. In particular, as Figure 2 illustrates, the share of municipalities that submitted plans for approval by the Commission within one year of the date of joining was higher in the UK where, however, there is a strong polarisation: municipalities tend either to be very active and present their SEAP within one year or to delay submission of the document for a long time (up to 6–7 years). In Italy, the majority of municipalities prepared their plans within 2–3 years from the date of joining and, as interviews explain [Interviews 1, 4, 9], the delay was mainly due to technical difficulties and a lack of resources.

With regard to the prioritisation of and the readiness to comply with the technical requirements for SEAPs, Italy leads: 83.7% of municipalities that joined the CoM have had their SEAPs approved by the COMO, while in the UK this percentage is only around 56% (see Table 2). It is worth stressing that most of the pending demands in Italy are dated between 2014 and 2017, while in the UK they are dated back to 2010–2014, apparently reflecting a rather low political priority for the municipalities concerned, in particular after 2014.

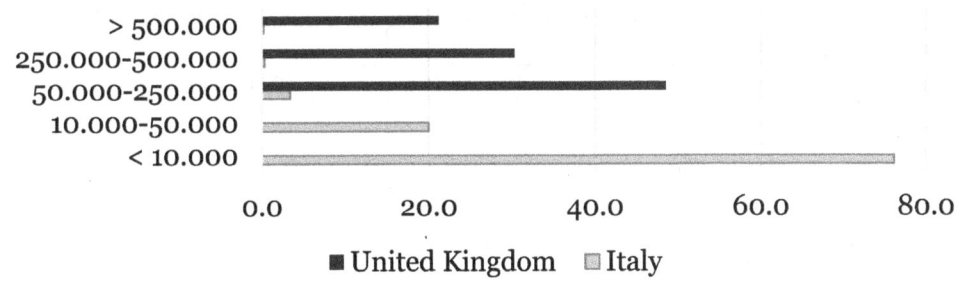

Figure 1. Municipalities participating in the CoM by population (%).
Source: Author's calculation.

Table 1. Implementation of monitoring in Italy and the UK as of January 2017.

	Italy			United Kingdom		
Population	N SEAPs approved	N of SEAPs with monitoring	Percentage of monitored SEAPs	N SEAPs Approved	N of SEAPs with monitoring	Percentage of monitored SEAPs
<10.000	1811	558	31%	0	0	0
10.000–50.000	456	176	39%	0	0	0
50.000–250.000	72	31	43%	7	3	43%
250.000–500.000	5	4	80%	7	4	57%
>500.000	5	4	80%	4	2	50%
Total	2349	773		18	9	

Source: Author's calculation.

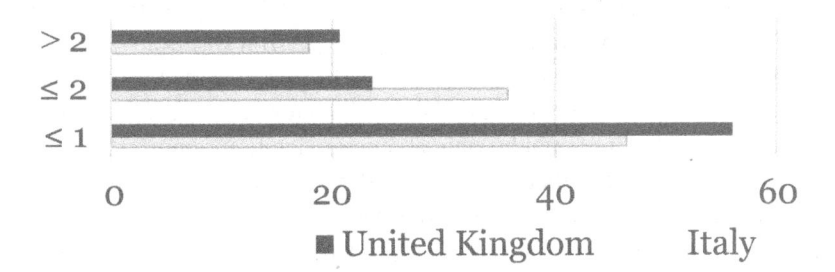

Figure 2. Time employed for the preparation of SEAPS (%).
Source: Author's calculation (using a sample of 500 Italian municipalities and 34 UK).

Concerning the target of CO2 reduction, in both countries there is a large group of municipalities that have established more ambitious objectives than those set by the EU Climate and Energy package 2020. Overall, as Table 3 shows, the reduction targets were higher in the UK than in Italy, with a slightly more numerous group of municipalities that sought to reach a 25–38% reduction, before the new target of 40% was introduced in 2016. At the same time, local strategies in both countries have, to a very limited extent, included adaptation measures.

Table 2. Status of commitment of municipalities participating in the CoM as of August 2017.

	Signatures	SEAPs submitted	SEAPs accepted	Pending
Italy	3268	3125	2716	196
UK	36	34	18	14

Source: Author's calculation.

Table 3. Targets of CO2 reduction defined by local SEAPs (as of August 2017).

Commitment	ITALY	UK
Mitigation(2020 target)	3027	29
Mitigation(2030)	26	0
Adaptation	0	0
Mitigation(2020 target) and Adaptation	46	4
Mitigation(2020, 2030) and Adaptation	26	1
Total	3125	34

Source: Author's calculation.

Survey respondents perceive the general relevance of actions performed within the CoM for the achievement of local objectives in the field of sustainable energy as rather high in both countries: 24/28 in Italy and 5/7 in the UK. In contrast, opinions split with regard to the CoM's usefulness for the development of local sustainable energy. The majority of Italian respondents consider the CoM to be helpful for improving specific aspects of their local policies for the reduction of GHG emissions (15/28); for others (8/28), the participation in the CoM has been fundamental and for a minority (5/28) it was nearly useless. In contrast, in the UK, 4 out of 7 consider the participation in the CoM useful for the overall strategy, but only 1 assessed it as helpful for some specific aspects and 2 out of 7 think that it has not brought about any significant improvement. The interviews also confirm that more respondents in Italy have perceived the practical relevance of the CoM for improving local strategies for sustainable energy as important [Interviews 1, 6, 7], whereas UK respondents view the impact of this programme as symbolic [10–12].

Similarly, significant differences between the countries emerge on monitoring mechanisms, according to which municipalities should monitor implemented actions and publish the related data on the CoM website within two years following the submission of the SEAPs. The data should reflect the state of implementation (ongoing, completed, not started) along with performance indicators concerning GHG emissions, energy consumption and local energy production. The monitoring activities should then be subsequently updated every two years, along with Inventory Monitoring Emissions, and data on the availability of technical expertise and resources. Hence, while in the UK 9 of 18 approved SEAPs (50%) have implemented monitoring, in Italy only 25.3% of SEAPs have complied, which is close to the EU average of 22.2%. Significantly, in both countries, large and very large municipalities perform better than small and medium-sized ones (Table 1).

The survey results help to explain these findings, as significant differences exist in perceptions of this instrument among municipalities. In the UK, 4 out of 7 respondents highlighted the necessity to improve and strengthen the monitoring protocol, while their Italian counterparts considered it adequate and sufficient for measuring the effectiveness of implemented actions, emphasising the lack of financial and knowledge resources (25/28) for its proper implementation.

In the same way, significant differences exist between the two countries in the extent to which the authorities employ CoM as a source of reliable information about costs, benefits and improvements of sustainable energy policies. For example, the share of municipalities contributing to the mechanism of benchmarking, which allows the Covenant members to share excellent local initiatives and endorse them as useful actions for other authorities to replicate (up to 3 for municipality), has been higher in the UK: 50% of those who had their SEAPs approved against 37% in Italy, although there is a huge difference in absolute numbers of benchmarks shared by Italian (2721) and UK (27) local authorities. Remarkably, while the UK benchmarks have been produced by larger cities, the highest number of benchmarks in Italy (1871) comes from small municipalities with less than 10,000 inhabitants.

As for participation in CoM training and capacity building activities, the survey reveals quite different trends too. Although municipalities in both countries tend to be more frequently involved in events dedicated to some specific issues of their own interest (13/28 in Italy and 3/7 in the UK) rather than acting as promoters (7/28 in Italy and 1 in the UK), almost one-third of Italian municipalities (8/28) frequently followed various events promoted by the CoM, while 2 UK respondents had never taken part.

With regard to the general importance of the CoM as a source of reliable information and policy knowledge, analysis of the frequency with which its members use the dedicated section of the website for consulting and sharing methodological documents, case studies and best practices shows that the number of uploaded documents as well as the number of 'reads', 'downloads' and 'comments' is somewhat limited (a maximum of 40). Additionally, the average number of participants in webinars, which are the main tools for building and diffusing knowledge and experience, appears to be rather low compared to the overall CoM membership.

Finally, survey respondents perceive mutual trust and the reputation of being a trustworthy reciprocator to be more important for UK participants, while for Italian municipalities this becomes relevant only for special occasions (events, initiatives). Similarly, UK municipalities unanimously see the CoM as an important tool for raising the visibility of local government efforts to combat climate change (5 out of 7), while in Italy this was true for only one-third of respondents.

The survey mentioned the availability of high-quality internal staff and political leadership as being among the most important factors for successful participation in the CoM in both countries, with 2o of 28 Italian and 4 of 7 UK respondents, and 24 of 28 Italian and 5 of 7 UK municipalities, assessing these qualities as 'very important'. Participation in the CoM has brought about important organisational changes in most Italian municipalities, where actors have introduced new structures or coordination mechanisms to follow the SEAP-related activities, while in the UK, municipalities have assigned additional functions to existing energy or environmental officers.

In sum, the implementation scenario of the CoM in the two domestic contexts offers several insights on this programme's potential to enhance local authority action within EU climate policies. Its impact has proved to be much more significant in Italy, where strategic and cognitive patterns of usage have prevailed. In addition to a high number of municipalities joining the programme to adjust their local sustainable energy policies to the EU climate objectives, there has also been an intense involvement in a wide range of specific learning and cooperation activities promoted by the CoM, including training and benchmarking. The implementation of the monitoring mechanism has been rather slow in Italy, mainly due to lack of internal expertise and financial resources. A completely different pattern of participation arises for the UK, where participation is amongst the lowest in the EU, and the symbolic and legitimating usages of this programme dominate; a small group of local authorities quickly aligned with CoM guidance and actively delivered the monitoring and benchmarking required by the programme, but they have employed the CoM learning resources to a limited extent. Some UK SEAPs established ambitious targets and called for the strengthening of the monitoring mechanism based on pre-existing local climate strategies, while others have joined in a symbolic manner without performing any substantial activities. Overall, the perception of the CoM as a valuable tool for increasing international visibility of local authorities and promoting their image as a reliable partner in the field of sustainable energy policies has appeared stronger in the UK than in Italy.

Importantly, alongside these differences, the analysis also revealed similarities. Local authorities perceive CoM objectives as important for their own achievements in the field of sustainable energy policies and appreciate the effort of establishing long-term joint actions in line with EU priorities. The majority of respondents in both countries believe that availability of internal technical expertise, coupled with strong political commitment, represents the most relevant variables for successful implementation. They also believe that a joint monitoring mechanism is feasible and important, and there is a significant leverage effect of the CoM in terms of access to additional financial resources and involvement in international cooperative networks based on EU funding (in particular Life+, ELENA, Horizon 2020).

What about domestic contexts?

An overview of additional domestic characteristics can helpfully complete the above analysis of the implementation of the CoM in the two countries. As mentioned, the majority of Italian signatories are small and medium-sized cities and the symbolic usage of the programme prevailed in the country in its early stage – a trip to Brussels for the signature of the plan was important for the visibility and political career of mayors of small and medium-sized towns [Interviews 1, 5, 7]. However, attitudes towards the CoM have gradually changed over time, as it has become clear that participation in the programme requires specific capacities and long-term commitments [Interviews 2, 3, 8]. Despite financial difficulties and lack of expertise, the opportunities that the programme offers in terms of new knowledge and additional resources (even indirect) have pushed dozens of Italian local administrations to join in order to satisfy the cognitive and pragmatic demands of public servants. Many Italian administrations have used the CoM to upgrade or develop their local energy plans established by the poorly implemented National Law 10/1991, in an attempt to overcome the significant lack of internal technical expertise on issues such as the creation of emissions inventories, the calculation of economic costs and the assessment of the environmental impact of policies and measures (Lumicisi 2013).

However, the learning output related to the implementation of the CoM appears limited in Italy, as the process of formulation and implementation of SEAPS has normally involved local public servants in charge of environmental and energy planning or external experts, with poor mechanisms for horizontal policy coordination with other sectoral units. Moreover, alterations in local governmental majorities have negatively influenced the speed and quality of the implementation of SEAPs, jeopardising the continuity of commitments [Interviews 1, 2, 5, 7]. At the same time, many respondents highlighted that the progress, continuity and consistency of municipal strategies strongly depend on the commitment and expertise of the administrative staff in charge of them, as well as on their capacity to obtain additional resources and build partnerships with economic stakeholders at local level. Survey respondents almost unanimously agreed that political leadership is the most important factor for successful participation in the CoM.

Overall, participation in the CoM appears to have positively affected general knowledge and awareness in local administrations about the relevance of energy efficiency and saving, especially in terms of cost reduction and new opportunities for local development policies, in particular with regard to private investments. Some respondents have highlighted how lack of coordination and the absence of common methodological guidance at national and, in many cases, regional level, created barriers to successful implementation of the programme, along with a drastic reduction of public funds and procedural burdens impeding collaboration with private companies [3–5]. In fact, in those regions or provinces (e.g. Abruzzi, Chieti and Emilia-Romagna) where coordination was guaranteed, the implementation dynamics were quite smooth due to specific expert support, intensive exchange between local administrations and facilitated access to funding. It comes as no surprise that the majority of 539 monitoring plans submitted by Italian municipalities came from these areas.

Finally, although Italy has the highest number of signatories and benchmarks (2439 out of 4347 collected by the CoM), it also has the highest number of desertions. From September 2012 to February 2013, 2145 signatories reduced to 2089 (56 expulsions in less than 12 months). Remarkably, notwithstanding considerable bottom-up activism and a growing number of initiatives for energy efficiency carried out at local level, central government has largely disregarded the value and potential of local efforts for reaching the targets set by the EU Climate and Energy 2020 package. The National Plan, approved in 2013 for reducing greenhouse gas emissions in non-ETS sectors, included no reference to local initiatives; the plan defines measures for the implementation of Decision 406/2009 ('Effort Sharing') (European Union 2009). Against this background, local authorities have widely perceived the CoM as a valuable and unique opportunity to acquire knowledge, develop and improve policies and obtain financial resources in an increasingly important policy area.

The implementation of the CoM in the UK has been much less successful and it appears striking considering that the operational logic underlying the CoM is highly compatible with the prevailing policy style in UK environmental policy-making, characterised by a strong commitment to cooperation, administrative discretion and technical specialisation, but this is consistent with the similarly reluctant implementation of NEPIs during previous decades (Wurzel *et al.* 2013). The following factors explain the main obstacles to the diffusion of the CoM in the UK.

First, in most UK cities the position of mayor is a ceremonial one with no real decision-making power so, naturally enough, the title of the initiative itself did not evoke much enthusiasm on the part of local governments. Moreover, the CoM partially overlapped with a national initiative for local climate action, the Nottingham Declaration Partnership on Climate Change, which was launched in 2000, upgraded in 2011, and supported by over 300 local authorities. Under these conditions, it has been difficult for local authorities to appreciate the added value of the CoM and to justify participation in it, especially as it requires additional workload and costs for municipal staff.

It is consequently unsurprising that legitimising usage emerged in a group of large cities, such as Nottingham, Leicester, Manchester and Newcastle-upon-Tyne, which joined the network at its earliest stage, building upon their own previous initiatives for sustainable urban development. Only a few elements of strategic usage of the CoM could be observed in municipalities with elected mayors, such as Greater Manchester or Bristol, which have employed the programme to increase their international visibility, obtain financing or develop cooperation networks by attracting EU funding from the ELENA facility, and Horizon 2020. A further explanation of the low relevance of the cognitive dimension in the UK can be derived from the fact that during the last decade the national government has taken a number of successful steps to promote adaptation and develop the capacity of local authorities and other public and private sector organisations to use climate science for adaptation planning and decision-making (Porter *et al.* 2014). Interestingly, since the subsequent downgrading of UK renewable energy policies, local authorities started to perceive the CoM as a way to keep low carbon issues high on local political agendas [Interviews 9, 12].

Conclusions

By using a conceptual and analytical framework that combines the insights provided by a polycentric governance perspective and the concept of usage, this study delivers evidence on the potential benefits and shortcomings of polycentric systems for governing climate change in the EU.

While the general validity of the CoM framework for enhancing local climate action appears to be confirmed, a significant variability in its concrete impact has emerged between and within the two countries, as well as between the different dimensions of activity (objectives and targets, monitoring, learning). More specifically, the empirical results show that, independently of context, local authorities consider the CoM a relevant tool for increasing awareness of local authorities about the need for coordinated climate action based on joint political commitment, shared objectives and mutual monitoring. At the same time, local authorities' performance varies a lot within different strands of activity of the network, depending on both contextual conditions (e.g. the existence of similar domestic policy programmes) and local authorities' motivations and capacities, including the availability of knowledge, human and financial resources, city size and political will.

In fact, the analysis has demonstrated that strategic and learning resources that the CoM provides have been more actively employed and become a driver for policy change at the local level in Italy regardless of unfavourable domestic conditions. Indeed, the CoM has filled gaps in the domestic climate regime, as Italian municipalities have extensively joined the programme and committed to its objectives by acquiring new policy knowledge and adjusting their strategies to common targets notwithstanding short-term costs. Interestingly, the Italian case has brought to light a noteworthy learning capacity of small and medium-sized cities that, according to previous studies (Kern 2010, Van der Heijden 2017), normally show insufficient ability to implement governance innovations in their jurisdictions. In contrast, legitimising and symbolic patterns of usage have prevailed in the UK, mainly due to the overlap of the CoM with existing domestic climate initiatives that enabled the development of local climate knowledge and strategies prior to the establishment of the CoM. These factors have determined higher local authority ambition on, for example, emission targets, and lower motivation to be involved in a system with comparatively lower standards (e.g. monitoring).

In addition to these findings, a larger-scale study covering 891 municipalities (CoM 2013) shows that the Covenant's perceived effectiveness in terms of raising local awareness about sustainable energy has been high (80% respondents), whereas satisfaction with the COMO and JRC supporting activities has been limited (around 45% are positive) and rarely used (between 40% and 45% of respondents have asked for it). Major criticisms concern limited linguistic capacity, poor timing in providing feedback and a lack of guidance, particularly for the implementation reports.

Thus, if it is true that the strength of new polycentric governance instruments lies in their flexibility, this analysis shows that their contribution to increasing the EU's problem-solving capacity may be further strengthened through a more careful calibration in view of the potential and concrete policy demands of target actors. Major efforts in this direction would respond to the two-fold challenge of better effectiveness and legitimation of the CoM vis-à-vis other domestic or TMN for climate in Europe.

To this end, the learning dimensions have yet to be studied in depth, as our results show that we cannot take for granted that the mechanisms of coordination and socialisation within networks provide for learning in at least the following two directions (Radaelli 2008): 'from the top', for example, through the system of common templates, methodologies and reporting, or learning 'from below' through benchmarking and joint training.

Paying attention to the micro-foundations of the policy process has proved helpful for spelling out the conditions under which policy actors are likely to employ opportunities and commit to collective climate action (Hongtao et al. 2017, Tosun and Schoenefeld 2017) supported by the EU. In this perspective, more quantitative analysis, covering a larger number of cases in different domestic contexts, would be needed, which should be necessarily complemented by qualitative research explaining how and why actors move between different policy levels and how 'networks that form among actors with fundamentally different goals matter and perform' (Woll and Jacquot 2010, p. 118).

The aforementioned aspects appear to be crucial for improving our understanding of political processes in the EU, considering the current challenge of internal political restructuring related to Brexit, along with growing external demands for stronger leadership in view of the withdrawal of the USA from the Paris commitments.

Acknowledgements

This work draws on the Research project 'Climate change policies and governance. The experience of EU voluntary instruments', supported by the Department of Political Science, Law and International Studies (2017–2018). Some preliminary ideas were presented to the ECPR Joint Session Workshop 'Whither the Environment in Europe? (Pisa, 24–28 April 2016), and further refined in a paper presented to the International Public Policy Association Conference (Singapore, 27–30 June 2017). I am particularly grateful to the Special Issue editors: Charlotte Burns, Andrea Lenschow and Anthony Zito, who organised the Workshop in Pisa and provided invaluable suggestions for improving this work. I also thank Elena Ceretta and Giuseppe Acconcia for their support in collecting data and conducting interviews, as well as three anonymous referees for insightful comments and constructive criticism.

Disclosure statement

No potential conflict of interest was reported by the author.

Funding

This work draws on the Research project 'Climate change policies and governance. The experience of EU voluntary instruments', supported by the Department of Political Science, Law and International Studies (2017-2018).

ORCID

Ekaterina Domorenok http://orcid.org/0000-0002-4791-1480

References

Abbott, K.W., 2012. The transnational regime complex for climate change. *Environment and Planning C Government and Policy*, 30 (4), 571–590. doi:10.1068/c11127

Berkhout, F., et al., 2015. European policy responses to climate change: progress on mainstreaming emissions reduction and adaptation. *Regional Environmental Change*, 15 (6), 949–959. doi:10.1007/s10113-015-0801-6

Bouteligier, S., 2013. *Cities, networks and global environmental governance*. New York: Routledge.

Bulkeley, H., et al., 2012. Governing climate change transnationally: assessing the evidence from a database of sixty initiatives. *Environment and Planning C: Government and Policy*, 30 (4), 591–612. doi:10.1068/c11126

Busch, H., 2015. Linked for action? An analysis of transnational municipal networks in Germany. *International Journal of Urban Sustainable Development*, 7, 213–231. doi:10.1080/19463138.2015.1057144

Covenant of Mayors (CoM), 2013. *Mid-term evaluation of the Covenant of Mayors* [online], Technopolis Group, Fondazione Eni Enrico Mattei, Hinicio and Ludwig-Bölkow-Systemtechnik. Final Report. Available from: http://www.technopolis-group.com/report/mid-term-evaluation-covenant-mayors-initiative

Covenant of Mayors (CoM), 2015. Press release. Available from: http://www.cove nantofmayors.eu/about/covenant-of-mayors_en.html

European Commission, 2010. *Communication Europe 2020 - a strategy for smart, sustainable and inclusive growth of 3 March 2010 [COM(2010) 2020] final* [online]. Available from: https://eur-lex.europa.eu/legal-content/en/ALL/?uri= CELEX%3A52010DC2020

European Commission, 2011. *Communication A budget for Europe 2020 of 29 June 2011 [COM(2011)500] final* [online]. Available from: http://ec.europa.eu/budget/library/ biblio/documents/fin_fwk1420/MFF_COM-2011-500_Part_I_en.pdf

European Union, 2009. Decision 406/2009/EC of the European parliament and of the Council of 23 April 2009 on the effort of member states to reduce their greenhouse gas emissions to meet the community's greenhouse gas emission reduction commitments up to 2020. *Official Journal of the European Communities* OJ L, 140, 136–148. 5 June.

European Union, 2012. Directive 2012/27/EU of the European parliament and of the Council of 25 October 2012 on energy efficiency, amending directives 2009/ 125/EC and 2010/30/EU and repealing directives 2004/8/EC and 2006/32/EC. *Official Journal of the European Communities* OJ L, 315, 1–56. 14 November.

Hakelberg, L., 2011. *Governing climate change by diffusion. Transnational municipal networks as catalysts of policy spread* [online]. Freie Universität Berlin. FFU Report ISSN 1612-3026. Available from: https://www.polsoz.fu-berlin.de/pol wiss/forschung/systeme/ffu/aktuelle-publikationen

Heidbreder, E.G., 2017. Strategies in multilevel policy implementation: moving beyond the limited focus on compliance. *Journal of European Public Policy*, 24 (9), 1367–1384. doi:10.1080/13501763.2017.1314540

Heidrich, et al., 2016. National climate policies across Europe and their impacts on cities strategies. *Journal of Environmental Management*, 16 (8), 36–45. doi:10.1016/j.jenvman.2015.11.043

Heyvaert, V., 2013. What's in a name? The Covenant of Mayors as transnational environmental regulation. *Review of European Community and International Law*, 22 (1), 78–90. doi:10.1111/reel.12009

Hildén, M., ed., 2017. Special issue: experimentation for climate change solution. *Journal of Cleaner Production*, 16 (9), 1–234. doi:10.1016/j.jclepro.2017.09.019

Hongtao, Y., Krause, R.M., and Feiock, R., 2017. Back-pedaling or continuing quietly? Assessing the impact of ICLEI membership termination on cities' sustainability actions. *Environmental Politics*, 26 (1), 138–160. doi:10.1080/ 09644016.2016.1244968

Jacquot, S. and Woll, C., 2003. Usage of European integration – Europeanisation from a sociological perspective. *European Integration Online Papers*, 7 (12), 1–12.

Joint Research Centre (JRC), 2015. *The covenant of mayors: in depth analysis of sustainable energy action plans* [online]. Policy Report. Available from: https:// www.covenantofmayors.eu/support/funding-instruments_bg.html [Accessed 20 February 2017].

Jordan, A. and Adelle, C., 2013. *Environmental policy in the EU. Actors, institutions and processes.* London: Routledge.

Jordan, A., Schout, A., and Zito, A. 2005. *Coordinating European Union environmental policy: shifting from passive to active coordination* [online]. CSERGE Working Paper EDM, No 04-05. Available from: http://hdl.handle.net/10419/80243

Jordan, A.J., et al., 2015. Emergence of polycentric climate governance and its future prospects. *Nature Climate Change*, 5, 977–982. doi:10.1038/nclimate2725

Jordan, A.J., et al., 2018. *Governing climate change. Polycentricity in action?* Cambridge: Cambridge University Press.

Kern, K., 2010. Climate governance in the EU multi-level system. *The role of cities, Paper presented at the Fifth Pan-European Conference on EU Politics*, 23–26 June, Porto.

Kern, K. and Bulkeley, H., 2009. Cities, Europeanization and multi-level governance: governing climate change through transnational municipal networks. *Journal of Common Market Studies*, 47 (2), 309–332. doi:10.1111/j.1468-5965.2009.00806.x

Lumicisi, A., 2013. *Le città come protagonisti della green economy*. Milano: Edizioni Ambiente.

Matschoss, K. and Repo, P., 2018. Governance experiments in climate action: empirical findings from the 28 European Union countries. *Environmental Politics*, 27 (4), 598–620. doi:10.1080/09644016.2018.1443743

Ostrom, E., 2009. *A polycentric approach for coping with climate change* [online]. The World Bank. Policy Research Working Paper. Available from: https://elibrary.worldbank.org/doi/abs/10.1596/1813-9450-5095

Ostrom, E., 2010. Polycentric systems for coping with collective action and global environmental change. *Global Environmental Change*, 20, 550–557. doi:10.1016/j.gloenvcha.2010.07.004

Ostrom, V., 1999. Polycentricity – part 1. *In*: M. McGinnis, ed., *Polycentricity and local public economies: readings from the worshop in political theory and policy analysis*. Ann Arbor: University of Michigan Press, 52–74.

Porter, J.J., Demeritt, D., and Dessai, S., 2014. *The right stuff? Informing adaptation to climate change in British local government* [online]. Sustainability Research Institute, University of Leeds. Paper No 76. Available from: http://www.see.leeds.ac.uk/fileadmin/Documents/research/sri/workingpapers/SRIPs-76.pdf

Radaelli, C., 2000. Policy transfer in the European Union: institutional isomorphism as a source of legitimacy. *Governance*, 13 (1), 25–43. doi:10.1111/gove.2000.13.issue-1

Radaelli, C., 2008. Europeanization, policy learning, and new modes of governance. *Journal of Comparative Policy Analysis*, 10 (3), 239–254. doi:10.1080/13876980802231008

Radaelli, C., Dente, B., and Dossi, S., 2012. Recasting institutionalism: institutional analysis and public policy. *European Political Science*, 11 (4), 537–550. doi:10.1057/eps.2012.1

Tosun, J. and Schoenefeld, J.J., 2017. Collective climate action and networked climate governance. *WIREs Climate Change*, 8, e440. doi:10.1002/wcc.440

Treib, O., 2014. Implementing and complying with EU governance outputs. *Living Reviews in European Governance*, 9 (1), 1–47. doi:10.12942/lreg-2014-1

Van der Heijden, J., 2017. *Voluntary programs for sustainable buildings and cities: opportunities and constraints in decarbonising the built environment*. Cambridge: Cambridge University Press.

Woll, C. and Jacquot, S., 2010. Using Europe: strategic action in multi-level politics. *Comparative European Politics*, 8 (1), 110–126. doi:10.1057/cep.2010.7

Wurzel, R.K.W., Zito, A.R., and Jordan, A.J., 2013. *Environmental governance in Europe. A comparative analysis of new environmental policy instruments*. Cheltenham: Edward Elgar.

Appendix

Interviews conducted.

Interview number	Municipality	Period
1	Bari	July 2017
2	Bologna	April 2016
3	Cernusco sul Naviglio	May 2015
4	Padova	May 2016
5	Palermo	July 2017
6	Pesaro	September 2017
7	Venice	June 2017
8	Verona	October 2015
9	Glasgow	June 2017
10	Durham	June 2017
11	Leeds	June 2017
12	Poole	March 2017

Compliance with EU environmental law. The iceberg is melting

Tanja A. Börzel and Aron Buzogány

ABSTRACT
The European Union (EU) has become the main driver for environmental policy output for its member states whose number has more than tripled over the past four decades. The EU's deepening and widening has led researchers to expect more non-compliance with EU environmental legislation. In fact, however, the implementation gap has narrowed over the past 25 years. Except for Southern enlargement, taking on new member states has not exacerbated the EU's compliance problem in the field of environmental policy. Nor has the expansion of the environmental *acquis*. This is explained by the European Commission's strategies of managing and enforcing compliance. EU environmental policy has become less demanding on member states since it increasingly tends to amend existing rather than set new legislation. Simultaneously, the Commission has developed new instruments to strengthen member state capacities to implement EU environmental legislation.

Introduction

Preventing member states from misusing national regulations as non-tariff barriers in the Internal Market was the main driver for developing a comprehensive body of environmental legislation (Zito 2000) and its 'journey to centre stage' (Haigh 2015). Becoming a proper European Union (EU) competence rather late, environmental policy has been one of the policy fields where European integration has substantially advanced during the last decades. The environmental *acquis* grew rapidly (Holzinger et al. 2006); the EU has become the main source of environmental policy output in the member states (Jordan and Adelle 2012; Delreux and Happaerts 2016). Simultaneously, EU environmental policy has suffered from serious compliance problems. It is the policy area with the second highest number of violations of EU law even without controlling for the legislation in force (see Hoffman 2019). The high level of non-compliance in this core European integration area has regularly fuelled concerns about a (growing) compliance problem in the EU (Collins and Earnshaw 1992; Jordan 1999; Haigh 2015).

The EU's growth is not limited to the environmental *acquis*; the number of member states has also tripled. A growing body of EU environmental law to enforce and an increasing number of states to monitor, support, and socialize should lead us to expect more non-compliance, particularly since the Commission's resources have not matched the growth in law and member states. However, if we control for the increased numbers of environmental laws and member states that could potentially violate them, non-compliance has fluctuated and overall declined since 1994. We argue that decreasing non-compliance is part of a long-lasting trend. Conditions for non-compliance today differ from those existing 40 years ago because of the changing nature of EU law. EU (environmental) law has become less demanding, tending to amend existing rather than introduce new legislation. Moreover, the Commission has developed a whole set of new instruments to strengthen the compliance capacity of (new) member states. Pre-accession conditionality, for instance, explains why, unlike Southern Enlargement in the 1980s, the accession of 12 new members in the 2000s has not caused a spike in non-compliance with EU (environmental) law.

The literature on member state compliance and implementation of EU environmental law is exceptionally rich (cf. Tosun 2012; Bondarouk and Mastenbroek 2018). We contribute a longitudinal perspective on patterns of non-compliance with EU environmental legislation that covers more than three decades. Focusing on EU-level developments when interpreting these time-variant patterns, we rely on two well-established approaches in EU compliance studies—enforcement and management (Tallberg 2002). These approaches provide different but not mutually exclusive perspectives that help explain the overall decline in breaches of EU environmental law and their temporal fluctuations.

We structure the contribution as follows. The first part briefly reviews research on (non-) compliance with EU environmental legislation, paying particular attention to how the literature has analyzed compliance and implementation over time. Using a dataset of infringement procedures launched by the European Commission against the member states between 1978 and 2016, we then present a longitudinal mapping of non-compliance with EU environmental law. We show that non-compliance has fluctuated but declined overall since the Internal Market's completion. The second section discusses the extent to which the compliance literature can account for the two major findings—fluctuation and decline—that our temporal analysis reveals. We conclude by discussing the policy implications of our main arguments and identifying new research avenues.

As time goes by: the implementation of EU environmental legislation

EU policies adopted in Brussels require legal implementation and practical application in the member states. In theory, EU law supersedes national regulations and entrenched practices. In practice, there is substantial cross-temporal, cross-national, and cross-policy variation in compliance with EU law. In past decades, the literature has sought to explain what most identified as a growing compliance problem in the EU (see especially Angelova et al. 2012; Treib 2014). Qualitative studies usually focus on policy practices within states that lead to the legal adoption (or non-adoption) of EU policies or their practical implementation (or non-implementation). In this context, the implementation of EU environmental policy was a particularly popular case in the late 1990s (Knill 1998; Knill and Lenschow 2000; Jordan 2001; Börzel 2003). These studies highlight the existence of a substantial implementation gap; they also contributed to the theoretical and conceptual understanding of the research agenda on compliance in the EU (Treib 2014). Nonetheless, researchers have seldom investigated variation across time. Rather, many studies start from the assumption that the EU is facing a compliance problem and seek to explain why this is so. Detailed case studies of different pieces of EU environmental law largely drove research (Bondarouk and Mastenbroek 2018). Scholars located the major causes of non-compliance at member state level. Accordingly, country-specific variables, such as legal culture and administrative traditions, as well as state power and state capacity, were the analytical focus accounting for differences in non-compliance. This was also the case when the accession of Greece, Portugal and Spain in the 1980s triggered a debate on whether the EU had acquired a 'Southern problem' (La Spina and Sciortino 1993; Pridham and Cini 1994). The three Southern newcomers had significant difficulties in complying with EU law, and particularly environmental policy. The literature suggested these problems stemmed from some common deficiencies the Mediterranean countries shared with regard to their administrative and political systems, the weakness of civil society, and low levels of socio-economic development (Spanou 1998; Eder and Kousis 2001; Börzel 2003; Koutalakis 2004).

The 1995 (EFTA) enlargement involved three new countries, two from the North (Sweden and Finland) and one from Central Europe (Austria). Their accession did not raise much concern regarding implementation and compliance, partly because analysts regarded these countries as 'environmental pioneers'. Instead, the EFTA enlargement triggered debates about how small states can become influential by 'uploading' their innovative policies to the EU-level (Lauber 1997; Kronsell 2002).

The accession of the 10 Central and Eastern European (CEE) countries (plus Malta and Cyprus) in 2004 and 2007 revived the debate about environmental laggards in the EU (Skjærseth and Wettestad 2007). Indeed, the CEE countries have shared many symptoms of the 'Mediterranean Syndrome': inefficient administrations ridden by patronage and corruption, legacies of authoritarianism, weakly organized societal interests, and lower levels of socio-economic development (Börzel 2009; Börzel and Buzogány 2010). Moreover, the environmental *acquis* had grown considerably, rendering its implementation costlier than when the three Southerners joined. Many students of EU environmental policy-making expected a slowdown or even set back as a result of Eastern Enlargement (Jehlicka and Tickle 2004; Liefferink et al. 2009).

Measuring non-compliance over time

How has non-compliance with EU environmental law fared over the past 25 years? Has the subsequent deepening and widening of the EU exacerbated its implementation gap? Has the Southern problem turned into an Eastern problem? To answer these questions, we trace non-compliance over time, making use of longitudinal data on infringements of EU environmental law.

Compliance and implementation research uses a large variety of different measurements to account for successful policy implementation. Most studies on compliance and implementation of EU environmental policies still rely on qualitative evaluation (see Bondarouk and Mastenbroek 2018). While such studies have the benefit of providing in-depth assessments of the implementation process, they are difficult to compare across member states or over time. Quantitative measurements are less fine-grained, but allow for more systematic comparisons. Many studies use quantitative data on the transposition of EU directives as a compliance measure (e.g. Steunenberg and Rhinard 2010). However, these data rely on notifications by member states referring only to the *timely* transposition of directives into national law; it does not cover the incorrect legal implementation of directives or the incorrect application of directives, regulations and treaty articles. Other studies use implementation reports prepared by consultancies or analyze the implementation measures in great detail (König and Mäder 2014; Zhelyazkova et al. 2017) but they only cover a limited number of member states and/or member states in a limited time period and selected policy sectors.

Infringement procedures provide a more comprehensive analysis of non-compliance with EU legislation (Börzel 2003). Art. 258 TFEU specifies that the European Commission can open infringement proceedings against member states in violation of EU law. It can base its action on citizen complaints, petitions and questions by the European Parliament, non-communication of the transposition of Directives or simply on its own initiative. The infringement proceedings consist of various stages, starting with a 'formal letter' and continuing with a 'reasoned opinion'. If member states fail to respond adequately to the Commission's inquiry, it can refer the case to the European Court of Justice (CJEU), which ultimately can impose a financial penalty.

Unlike the other quantitative measures of non-compliance mentioned, infringement proceedings have the advantage of covering all types of legal acts and possible violations over a long period. While the Commission is neither capable nor willing to legally pursue all violations it detects and therefore consistently focus on cases of systemic and persistent non-compliance, previous studies have found no evidence of a bias in infringements towards certain member states or policy sectors (Börzel et al. 2010; Börzel forthcoming). Nevertheless, infringement proceedings are certainly no measure for the absolute level of non-compliance. Scholars have rightly criticized them for representing merely the 'tip of the iceberg', the visible cases of non-compliance (Hartlapp and Falkner 2009). Additionally, from a legal perspective we cannot yet regard infringement procedures as breaches of law as they refer to cases where the European Commission has good reasons to suspect non-compliance (Smith 2016). Legal proof requires a conviction by the CJEU, which sides in more than 90 percent of cases with the Commission against the member states. These objections notwithstanding, infringement proceedings provide unbiased insights into member state non-compliance with core areas of the *acquis communautaire*. They remain the most systematic and comparable information source on non-compliance, allowing us to trace variance across member states, policy sectors, and time.

We use the Berlin Infringement Database (cf. Börzel and Knoll 2012; Börzel forthcoming) which includes 2,341 infringements cases the Commission brought against member states between 1978 and 2016 in the environmental policy field. Unlike the data published in the European Commission's Annual Reports on Monitoring the Application of Community Law or on DG Environment's 'Legal Enforcement' site, this dataset contains more detailed information regarding the nature of non-compliance, the type of law infringed on, the violating member state, and the measures EU institutions have taken to tackle non-compliance. We use 'reasoned opinions' as our main measure for non-compliance for two reasons. First, for the first stage of the infringement proceedings, the formal letter of warning, the Commission only provides aggregate data on the total number of cases brought against individual member states—as it considers information on individual cases confidential. Second, reasoned opinions concern the more serious non-compliance cases as they refer to issues that could not be solved at the previous, unofficial stages. Note that more than two-thirds of all the cases in which the Commission sends a warning letter are settled before it officially opens proceedings by issuing a reasoned opinion. To support our finding on the decline of non-compliance, we compare the trend in reasoned opinions with CJEU decisions, which represent the most contentious (and also more politicized) enforcement phase (Panke 2010).

Temporal patterns of non-compliance

As mentioned above, infringements may only capture the 'tip of the iceberg' (Hartlapp and Falkner 2009). We have no means to measure how large the iceberg really is. We can, however, assess whether the visible part of the iceberg has changed size over time. Simply comparing the number of infringement proceedings across time does not say much about changes in the level of non-compliance in the EU. The number of environmental legal acts in force has increased almost six-fold since 1978; 19 more member states that can potentially violate these acts have joined the EU. We measure reasoned opinion against the number of EU environmental laws in force multiplied by the member states in a given year that could potentially infringe them. Compare Figures 1 and 2 to see what difference it makes when we control for these 'violative opportunities' (Börzel et al. 2010; Börzel forthcoming). Figure 1 shows the overall distribution of reasoned opinions per year in 1978–2016. Numbers gradually increased until 2005 and then started to drop—despite 13 more member states joining. Figure 2 depicts reasoned opinions against the legislation in force and the number of member states in a given year. If we calculate infringements as the percentage of violative opportunities, non-compliance with EU environmental legislation already started to decline in the mid-1990s.

The implementation gap in environmental policy has narrowed rather than widened since 1994. However, we observe considerable fluctuation. While enlargements have not reversed the declining trend in non-compliance, they may account for temporary spikes. Figure 3 reproduces Figure 2 but groups the average reasoned opinions relative to 'violative opportunities' by member states joining the EU in the same year.

Greece, which joined in 1981, clearly drove the first spike in 1983/4. Note that there is on average a two-year lag between the occurrence of a violation and the Commission sending a reasoned opinion seeking redress. Accordingly, the effect of Portugal's and Spain's accession in 1986 started to show in 1988, and again, in 1993–95, when the period of grace the Commission had granted them expired (Börzel 2001). Nonetheless, the other member states also saw a significant increase in non-compliance after the Single European Act entered into force 1987, indicating factors other than Southern enlargement. While relative numbers started to drop in the second half of the 1990s, they flared up in 1998, in 2001, 2005, and

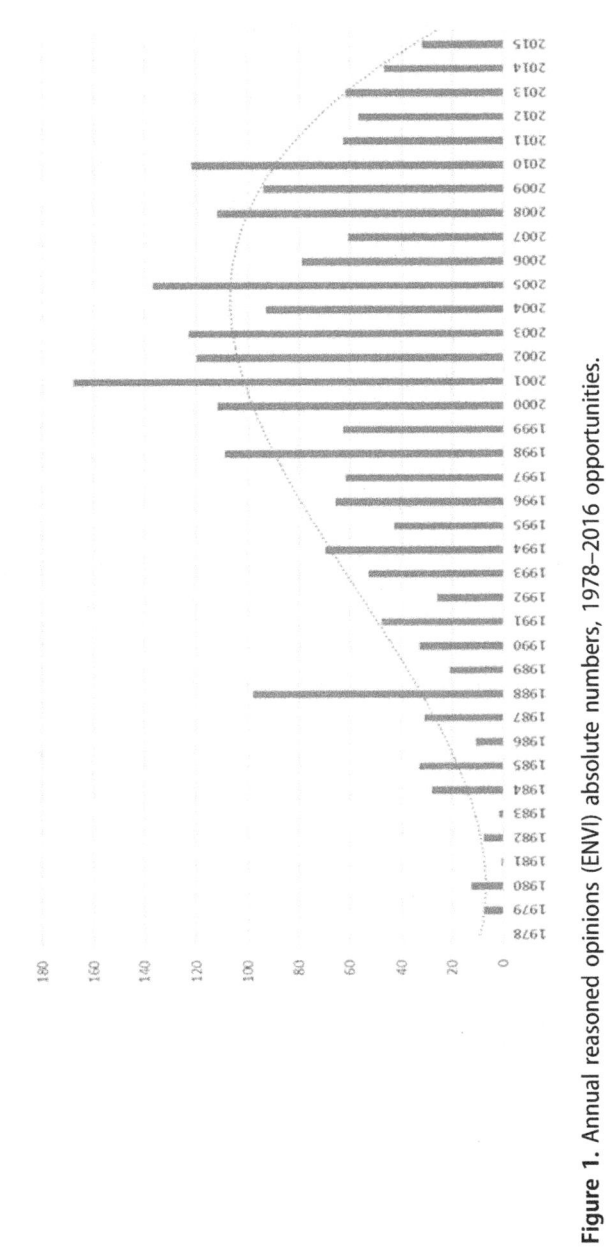

Figure 1. Annual reasoned opinions (ENVI) absolute numbers, 1978–2016 opportunities.
Source: Authors' own compilation with data from the Berlin Infringement Database.

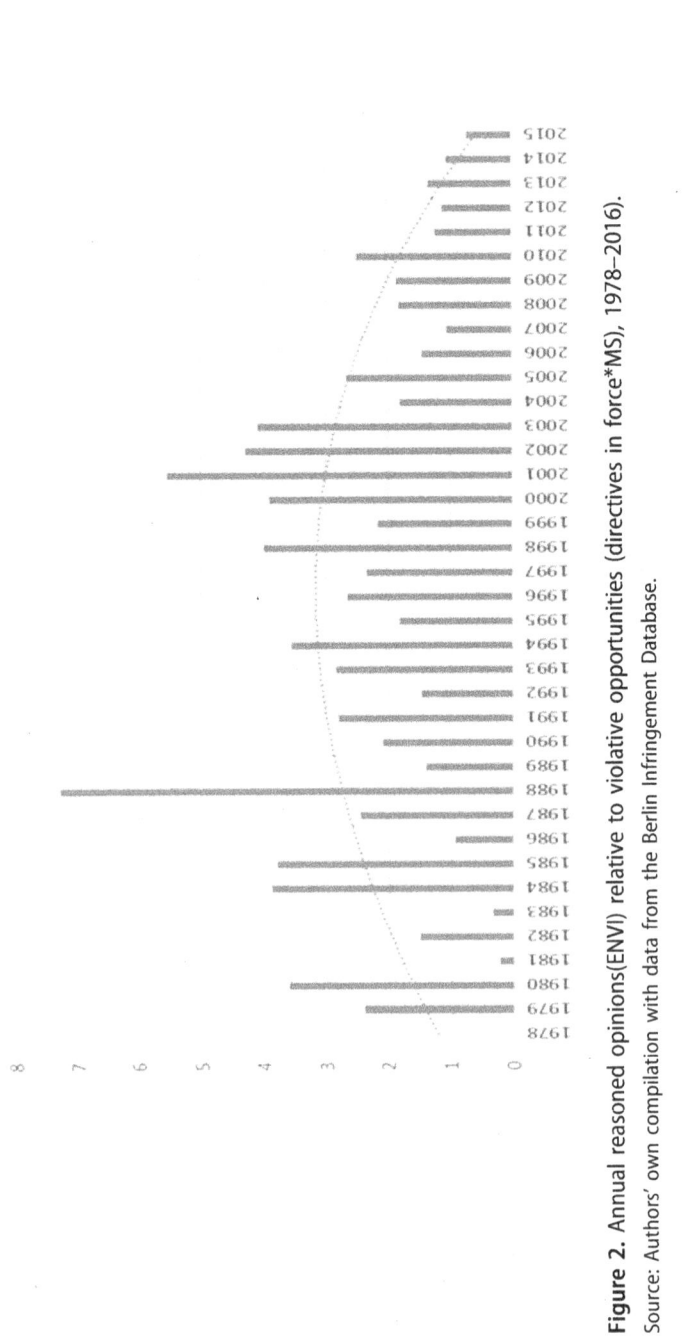

Figure 2. Annual reasoned opinions(ENVI) relative to violative opportunities (directives in force*MS), 1978–2016).
Source: Authors' own compilation with data from the Berlin Infringement Database.

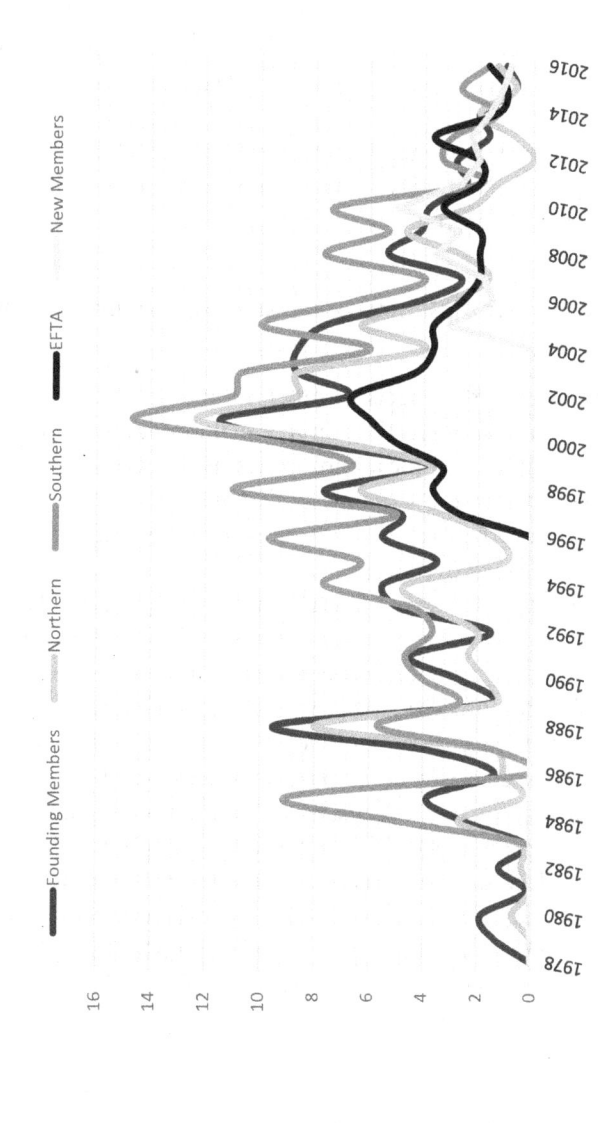

Figure 3. (Average) Annual reasoned opinions relative to violative opportunities in the area of environmental policy per member state, group means.
Source: Authors' own compilation with data from the Berlin Infringement Database.

2010. Three of the four spikes coincide with the accession of new member states. However, numbers increased for all member states, not only for the newcomers. Figure 4 illustrates that these trends are not specific to reasoned opinions but also show in later stages of infringement proceedings (see also Krämer 2006). Referrals to the Court and decisions of the Court concern breaches of EU law, which previous stages of the infringement proceedings could not resolve. As these referrals follow-up on reasoned opinions, there is a time lag between the spikes in Figures 3 and 4.[1] The trends remain the same: while there is fluctuation, we witness a clear downward trend since the mid-2000s.

In sum, the various enlargement rounds certainly affected the trajectory of EU non-compliance, but they cannot fully account for the periodic spikes. Thus the downward trend is counter-intuitive.

Understanding non-compliance: stricter enforcement and better management

Apart from debates over the effects of different enlargements, compliance and implementation research has been largely silent on longitudinal change. However, the two major theoretical approaches, which dominate the compliance literature and that the Commission's compliance strategies reflect (Tallberg 2002; Börzel 2003), offer some arguments for understanding the decline in non-compliance over time.

The *enforcement approach* understands compliance as a result of a rational choice (Downs *et al.* 1996). States weigh the costs of compliance against its benefits. The greater the distance between their policy preferences and the policies they have to comply with, the greater their compliance costs and the greater their incentive to defect. Punitive sanctions can inflict substantial non-compliance costs, deterring non-compliance (Fearon 1998). The weaker the monitoring and enforcement capacity of international institutions, the more likely is non-compliance. The *management approach*, in contrast, assumes that states lack the capacity rather than the willingness to manage compliance costs (Chayes and Chayes 1993; Raustiala and Slaughter 2002). Instead of monitoring and sanctioning, international institutions should help states cope with compliance costs by easing the burden and by strengthening their legal and administrative capacities.

The two dominant compliance theories allow the formulation of different arguments about why an increase or decline in non-compliance might occur over time in the EU. Enforcement approaches point to changes in the monitoring and sanctioning capacities of the European Commission increasing costs of non-compliance. For the management school, non-compliance depends on how effective the Commission is in managing compliance by clarifying behavioral requirements and assisting member states in coping

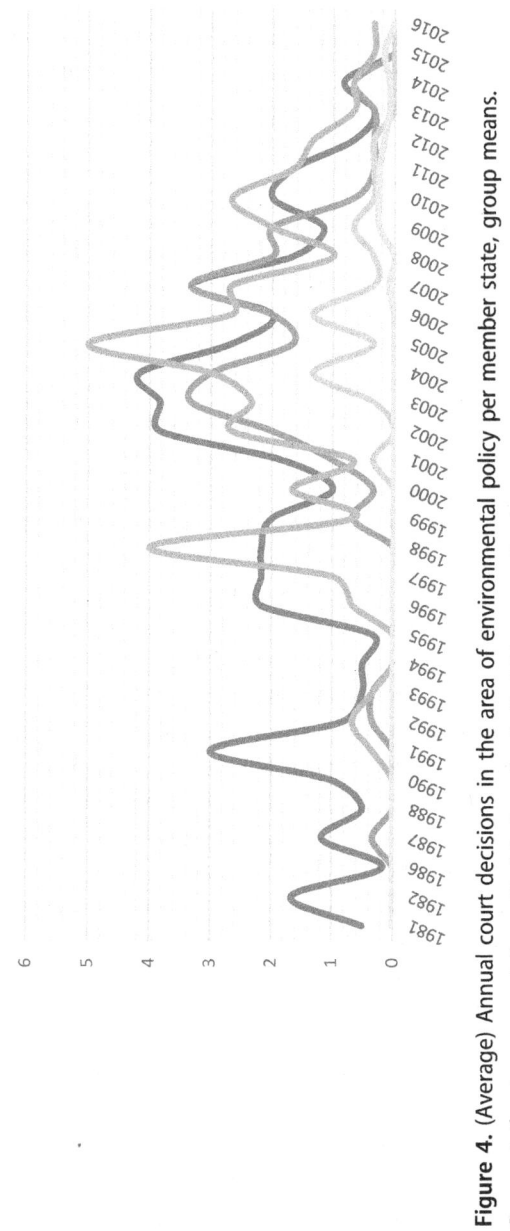

Figure 4. (Average) Annual court decisions in the area of environmental policy per member state, group means.
Source: Authors' own compilation with data from the Berlin Infringement Database.

with compliance costs. We do not systematically test these alternative accounts of declining non-compliance by using country- or sector-related variables, such as veto players, voting power, or administrative capacity, which are so prominent in the compliance literature (for an overview see Toshkov 2010; Angelova et al. 2012). While we acknowledge the importance of the domestic level for explaining non-compliance, we deliberately focus our attention on EU-level developments concerning environmental policy that are central to this volume (see Zito et al. 2019). We now proceed to offer a congruence analysis (George and Bennett 2005, p. 181–204) that matches changes in the enforcement and management capacities of the EU with changes in non-compliance over time.

Enforcement

Enforcement does not account for the overall downward trend but contributes to explaining fluctuations in the downward trend in non-compliance. Intensified enforcement efforts of the Commission often relate to an (upcoming) enlargement and largely drive the seven periodic spikes (see Figure 3). The 1984/85 peak was the combined effect of Greece's accession and the publication of the Commission's first annual report on the implementation of EU law in 1984 (Börzel 2001). The accession of Portugal and Spain and the Commission's more aggressive enforcement strategy to ensure the effective implementation of the Internal Market drove the second peak in 1988 (Tallberg 2002). The Market's official completion scheduled for 1992 also explains the third peak in 1994. The significant increase in infringements in the EU in 1998 related less to the 1995 EFTA enlargement and more to a 1996 internal reform of the infringement proceedings to improve effectiveness. To accelerate the process, the Commission decided to issue letters more rapidly than before. The number of letters the Commission sent increased significantly after the reform's implementation, which also stimulated the numbers of reasoned opinions (Börzel 2001). The second highest peak in 2001 reflects again a robust Commission effort to get the house in order before the 'big bang' enlargement. Likewise, the last two flare-ups in 2005 and 2010 relate to the Commission's strategy to level the playing field between new and old member states and counteract concerns about an 'Eastern problem' (Börzel 2009).

Monitoring and sanctioning non-compliance is contingent on resources. While 'violative opportunities' increased over time, the Commission's enforcement capacities did not. Contrary to its public image (see e.g. Moravcsik 2001), Brussels' bureaucracy has always been comparatively small, equaling the administration size of a European city such as Cologne. Still, the Commission has different tools available to monitor and sanction non-compliance.

With regard to monitoring, the Commission may launch its *own investigations*. With its limited resources, however, it relies heavily on decentralized monitoring information provided by citizen and business *complaints, parliamentary questions* and *petitions* (suspected infringements), and information member states provide concerning the transposition of directives into national law (referred to as *non-notifications* or *non-communication*).

The consistency and availability of information on suspected infringements varies significantly; we cannot break the data down by policy sectors. Figures 5 and 6 survey the years 1988 to 2010 for which data are consistently available. The Commission used to launch between 200 and 300 *investigations* per year—with the exception of the late 1980s, where the numbers were three times as high, probably due to the Commission's intensified effort to enforce EU law to complete the Internal Market. The numbers increased again post-Eastern enlargement but quickly returned to previous levels and have been dropping to an overall low in 2010. This may also relate to the introduction of new instruments such as SOLVIT and EU Pilot, designed to resolve compliance problems without resorting to infringement proceedings; they provide the Commission with information on potential non-compliance cases, reducing the need for launching own investigations.

Parliamentary questions and *petitions* have been much more limited but also peaked around the Internal Market's completion and the introduction of the Political and Economic and Monetary Union in the first half of the 1990s. They briefly flared up in 1991, probably also relating to Internal Market completion, and again around Eastern enlargement (2002–2004) and have declined ever since. *Complaints* steadily increased until the early 1990s, then started to drop but rose again in the mid-1990s to an overall high in 2002. Afterwards, numbers have continuously declined, particularly after 2004. Again, this may be due to SOLVIT and EU Pilot (Koops 2011, pp.180–181). Both offer alternative venues for business, societal organizations, and citizens to articulate their grievances about non-compliance.

Non-notification (non-communication) patterns, finally, are more diverse; enlargement effects appear largely to drive them. Numbers were high in 1996, after Austria, Finland, and Sweden joined, sky-rocketed in 2004, after the EU admitted 10 new members, and peaked once more in 2007 when Bulgaria and Romania joined.

Overall, monitoring information fluctuates significantly. There is no linear upward or downward trend in investigations, complaints, petitions, parliamentary questions, and non-notifications that would match the overall decline in infringements.

While the Commission heavily relies on decentralized mechanisms to monitor non-compliance, the EU centralizes sanctioning in the Commission's power to initiate infringement proceedings and the CJEU to impose financial penalties. Infringement proceedings provide a means to increase non-compliance costs by naming and shaming member states.

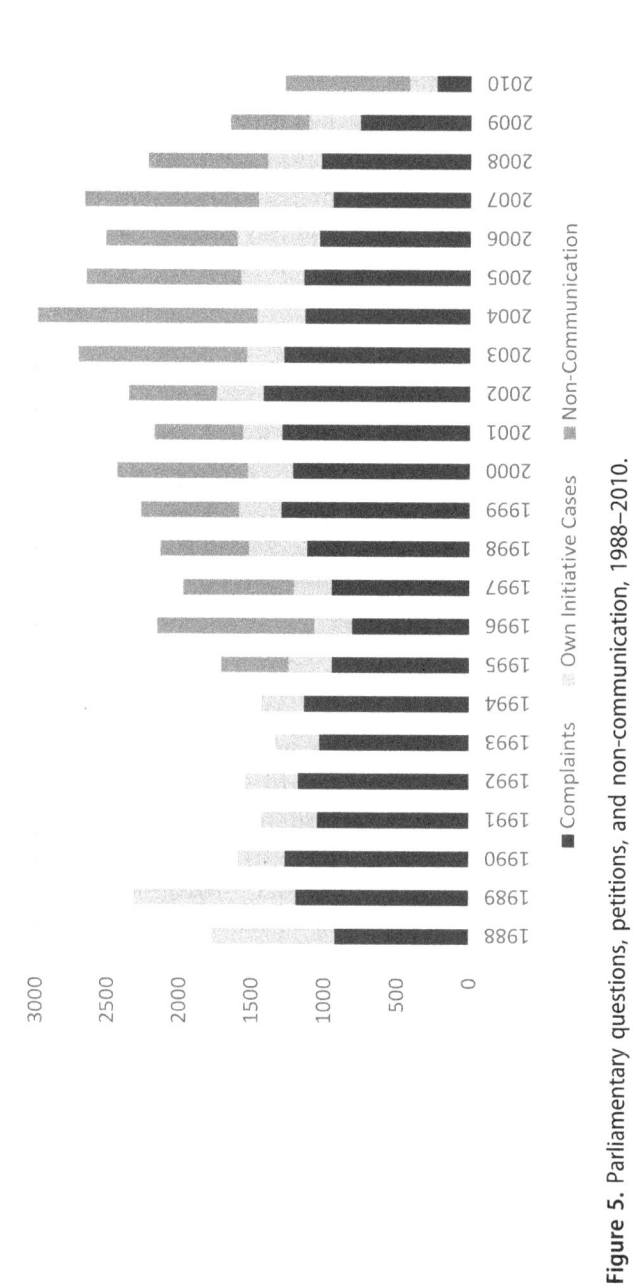

Figure 5. Parliamentary questions, petitions, and non-communication, 1988–2010.
Source: Authors' own compilation with data from Annual Reports 1989–2011.

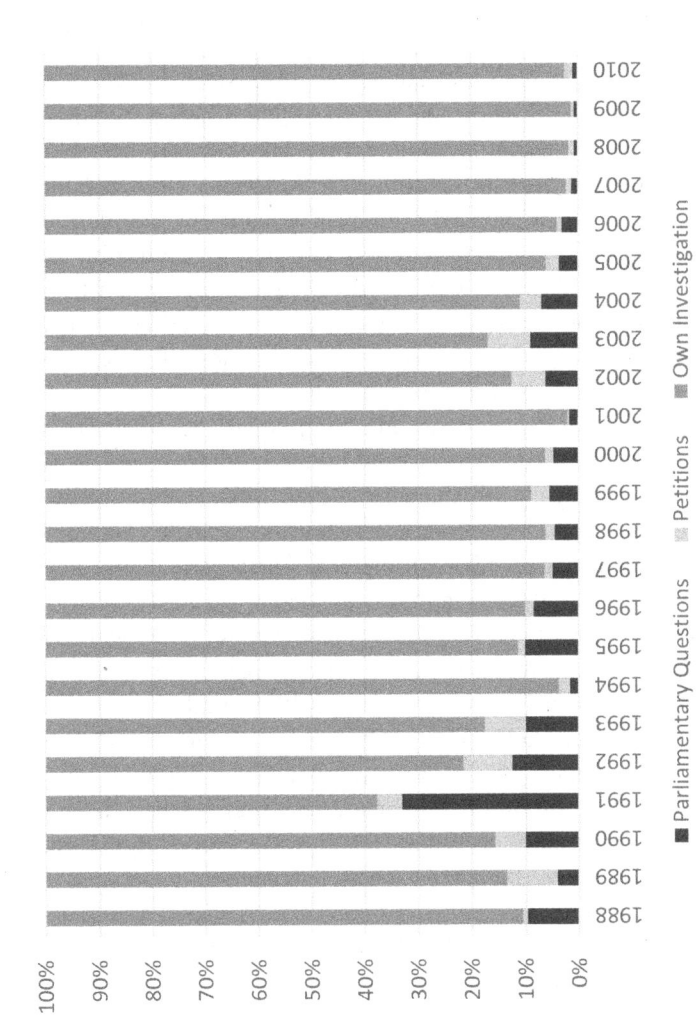

Figure 6. Complaints, own initiatives and own investigations, 1988–2010.

Source: Authors' own compilation with data from Annual Reports 1989–2011.

They also offer the road to financial sanctions. The Maastricht Treaty introduced the possibility of imposing *financial penalties* on member states that failed to comply with CJEU judgments (Article 260 TFEU). Article 260 became effective in 1993, just when infringement numbers relative to 'violative opportunities' had started declining. The CJEU invoked it first in 2000 when Commission had started a procedure against Greece in 1997 for not taking measures against the disposal of toxic and dangerous waste into the Kouroupitos River in Crete (Case C-387/97). It is questionable whether the mere anticipation of financial sanctions started to bring infringements down seven years before the member states learned that the CJEU was prepared to impose them. In 2009, Article 260 (2) of the Lisbon Treaty removed the necessity for the Commission to send a reasoned opinion before asking the CJEU to impose a financial penalty for non-compliance with its ruling. This may accelerate the sanctioning procedure by eight to 18 months (Peers 2012). Moreover, Article 260 (3) introduced a fast-track procedure allowing the Commission to ask, without a ruling of the CJEU under Article 258, the CJEU to impose financial sanctions if a member state has not reported the transposition of a directive.

Another mechanism for naming and shaming is the *Internal Market Scoreboard*, which the Commission established in 1997. Twice a year, it reports on member states' performance and their progress in implementing Single Market directives. The scoreboard allows for direct comparison of member state performance, highlighting the worst performers, not only among fellow governments but also in the public media (Tallberg 2002, p. 63). Only starting in 1997, has the Internal Market Scoreboard at best reinforced the downward trend. Cases of non-notification in this sector had dropped before 1997 and started to rise in 1998 until they reached an overall high in 2004 and 2007. Cases of incomplete and incorrect transposition and incorrect application of directives reached a high in 1995 after which they dropped; they climbed up again until they reached their overall high in 2006 before steadily declining. The introduction of the Scoreboard is unlikely to have driven these roller-coaster dynamics.

In sum, changes in the Commission's enforcement capacity help account for the fluctuations in non-compliance patterns over time. They, however, cannot explain the overall decline.

Management

Management theories highlight the importance of international institutions managing rather than enforcing compliance. Over the past 30 years, changes in EU environmental law and the Commission's focus on capacity building have helped member states cope with compliance costs. Some appear more effective in reducing non-compliance than others. Increasing use of amending legislation and financial and technical assistance to accession countries most closely relate to the declining trend. Amending legislation also reduces the need for enforcement due to lower compliance costs for all member states. Capacity building, in contrast, manages compliance in member states with weak capacities, which characterizes the vast majority of countries that have joined the EU in the past 40 years.

Compliance is particularly demanding on the legal and administrative capacities of member states when they have to introduce new laws. EU legal acts that require adaptation of existing domestic legislation incur lower compliance costs because they produce lower misfit. While the compliance literature contests the causal relevance of misfit (see e.g. Duina 1997; Börzel and Risse 2003; but see Falkner *et al*. 2004), several studies have shown that amending directives are less likely to give rise to delayed transposition than directives enacting new provisions (Mastenbroek 2003; König and Luetgert 2009; Haverland *et al*. 2011).

Data on new versus amending EU legislation is only available until 2012 and refers to directives. Nevertheless, Figure 7 shows a clear trend of the EU increasingly adopting amending directives; new legislation started to decline since 1994. Internal Market completion in the first half of the 1990s did more than reduce the need for new legislation. The rise of subsidiarity (Nugent 2016) and the shift of the policy agenda towards more controversial issues related to the extent to which the completed Internal Market should be regulated (Hix 2008) made it increasingly difficult for the Commission to table proposals for new laws. Since 2005, the EU has adopted more amending than new legislation. With the financial crisis, the EU appears reluctant to adopt legislation in several environmental policy subfields (Steinebach and Knill 2017).

Besides EU law becoming less demanding, the EU has also increased its efforts to build member state capacities for achieving compliance. First, it provides financial and technical assistance under various *EU funds* and *funding programs*. The Cohesion Fund and Community programs, such as the Action for the Protection of the Environment in the Mediterranean Region (MEDSPA), the Regional Action Programme on the Initiative of the Commission Concerning the Environment (ENVIREG), or the Financial Instrument for the Environment (LIFE), offer(ed) funding to assist member state's compliance with EU environmental legislation. Similarly, the EU established pre-accession funding schemes in the Eastern enlargement process, supplying Central and Eastern European candidate countries with significant financial and technical assistance (cf. Sissenich 2007, pp. 54–57). The EU organized technical assistance through 'twinning' programs and TAIEX, the EU's Technical Assistance Information Exchange Office (Börzel and Buzogány 2010). Member state experts have assisted candidate states in developing the legal and administrative structures required to

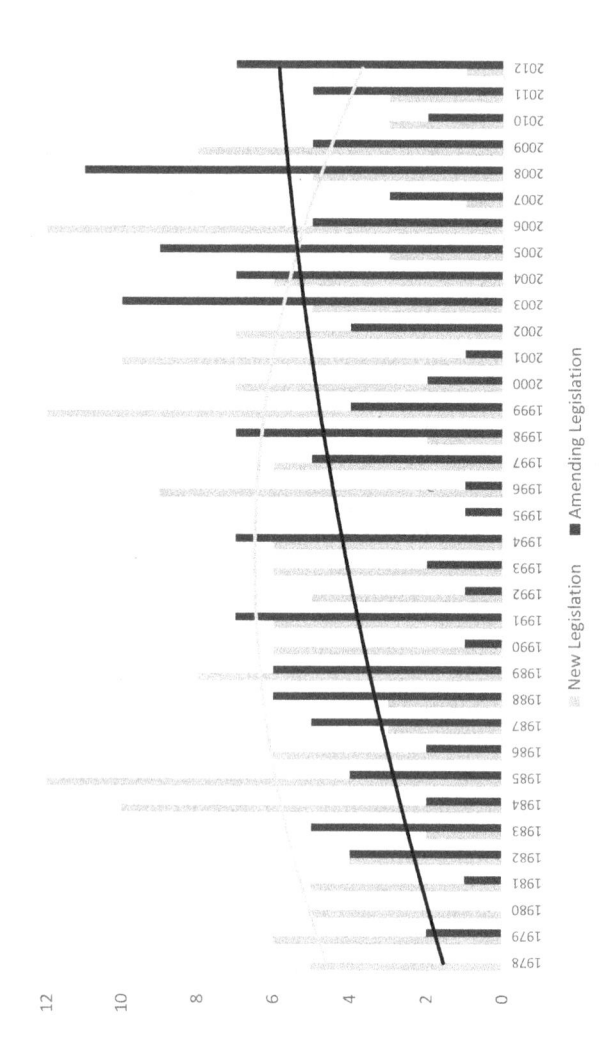

Figure 7. New and amending legislation in EU environmental policy, 1978–2012.

Source: Authors' own compilation using data obtained from the dataset compiled by Dimiter Toshkov, 'Legislative production in the EU, 1967-2012' http://www.dimiter.eu/Data.html, last access 23 March 2014.

implement effectively selected parts of EU environmental legislation. Transgovernmental networks between national administrators responsible for implementing EU law have fostered a common understanding of what compliance entails and facilitated processes of mutual learning about best practice. One of the oldest and most effective networks is the EU Network for the Implementation and Enforcement of Environmental Law (IMPEL). EU member states established it in 1992 as an informal network of European regulators and authorities concerned with the implementation and enforcement of environmental law. Likewise, EUFJE (European Union Forum of Judges for the Environment) promotes the enforcement of national, European and international environmental law by exchanging experiences on environmental case law and training judges.

Country studies provide ample evidence on how EU capacity building has helped accession countries and (new) member states improve their compliance with EU environmental law (Buzogány 2009). Pre- and post-accession financial instruments and twinning programs have played a major role in bringing new member states into compliance and may explain why Eastern enlargement has not exacerbated the EU's compliance problems (Schimmelfennig and Sedelmeier 2004; Börzel 2009, forthcoming). The combined effect of less (new) legislation and continued efforts at strengthening member state capacities to comply with existing legislation explains convincingly the narrowing environmental policy implementation gap.

Declining non-compliance: managing rather than enforcing the treaties

While enforcement approaches help explain the fluctuation in non-compliance, management approaches appear to largely explain the declining trend in non-compliance in the EU since the mid-1990s. Compliance with EU environmental law has become less demanding with the increasing adoption of amending rather than new legislation. This eases compliance costs, reducing the need for both enforcement and management. Nonetheless, the vast majority of member states that have joined the EU in recent decades suffer from weak capacities. They still face difficulties complying with EU laws that member states with higher regulatory standards and capacities have uploaded to the EU level (Börzel 2003, 2009). Therefore, the Commission has developed a comprehensive toolbox to strengthen member states' compliance capacities. Compliance research reveals that member states' administrative capacity is a powerful factor in explaining why some states comply less than others (Mbaye 2001; Hille and Knill 2006; Börzel *et al.* 2010; Börzel forthcoming). Accordingly, the Commission's use of pre-accession conditionality and pre-accession assistance to CEE candidate countries explains why they perform better than their Southern counterparts despite equally low administrative capacities (Börzel and Sedelmeier 2017). The Commission supported CEE (unlike Greece, Spain, and Portugal) in building capacities necessary to comply with EU law (Bruszt and Vukov 2017). While CEE states share bureaucratic similarities with Southern member states, general capacity indicators do not capture the specific capacities for implementing EU law. Qualitative environmental policy studies find significant implementation problems in the region (Orru and Rothstein 2015) and show that in some fields, such as climate policy, CEE countries are indeed among the brakemen (Braun 2014). Nevertheless, there is also sufficient evidence of conflicts related to the implementation of community legislation (Buzogány 2015; Sotirov *et al.* 2015) or of the empowerment of pro-compliance constituencies (Andonova and Tuta 2014, Dimitrova and Buzogány 2014) suggesting that the Commission's management strategy produced more than 'empty shells' (Dimitrova 2010) in a 'world of dead letters' (Falkner *et al.* 2008).

The Commission's management strategy has helped narrow the implementation gap. Capacity-building and easing compliance costs have proven more effective than surveillance and punishment (Bieber and Maiani 2014). Unsurprisingly, the Commission has developed a clear preference for management over enforcement (Hartlapp 2007; European Commission 2016). Reducing compliance costs by adopting less demanding legislation and strengthening member state capacities to cope with compliance obligations have reduced the need to bring legal action against member states. Yet, noncompliance is not only a matter of capacity. Enforcement remains an important element in the Commission's compliance approach, which often uses it in combination with management (Hoffman 2019). Powerful member states, for instance, tend to delay compliance with EU law when they anticipate domestic opposition (Börzel *et al.* 2010; Börzel forthcoming). There is still a substantial, albeit decreasing, number of systematic and persistent violations of EU law, e.g. the Fauna, Flora, Habitat Directive, which the Commission seeks to redress through punitive action (Gerdes *et al.* 2015). Thus, enforcement efforts contribute to but do not drive the downward trend.

Conclusion

The implementation gap in EU environmental policy has narrowed over the past 25 years—despite the tripling of member states that must comply with a four times larger environmental *acquis*. Except for the Southern enlargement, taking on new member states has not exacerbated the EU's compliance problem. Most importantly, Eastern enlargement has not made a difference, largely because of the Commission's management strategy for building capacity in the accession process. In response to the (anticipated) accession of 12 new members with weak capacity, the Commission prioritized the provision of financial and technical assistance to promote effective EU environmental legislation implementation. Next to capacity building, non-compliance has been declining because EU environmental policy has become less demanding by amending existing rather than creating new legislation.

Our findings have implications for future EU environmental policy. First, there is no contradiction between deepening and widening, at least when it comes to compliance. This is, second, because compliance is primarily a matter of administrative capacity rather than political willingness. A 'centralized Community inspectorate', as the Special Issue 'Green Dimension for the European Community' discussed 25 years ago, is not only '[...] at present politically unrealistic, if not possibly inappropriate' (Collins and Earnshaw 1992, p. 213); it is unlikely to make a substantial difference. Strengthening and harmonizing implementation activities in member states—something actors operating in the context of EU environmental policy have increasingly discussed (and practised) in recent years (Angelov and Cashman 2015; Hedemann-Robinson 2016; Knill *et al.* 2018) —appear much more promising.

We close by highlighting two major avenues for further research on compliance with EU (environmental) legislation. We have provided a first quantitative overview, which reveals an important trend and some plausible theoretical interpretations of the observed variation but calls for further empirical exploration and theoretical refinement. First, most of the environmental sector research has examined predominantly market-correcting policies setting production and product standards to fight environmental pollution (Article 191 TFEU). Over time, EU environmental law has become highly diverse in terms of policy sectors, policy tools and governance approaches. We have also seen an increase in the relevance of issue areas, such as climate change, energy or transport, which are often closely related to environmental policy. Moreover, we know that EU legislation differs significantly concerning clarity, novelty (versus amendments, see above), comprehensiveness, consistency, and practical recommendations of individual legal acts (Haigh 2015). Nevertheless, we know little about whether there are structural differences in how member states comply with different types of environmental or environment-related legislation – or whether there are differences in how the Commission guards the Treaty (Čavoški 2015). Future studies could differentiate between compliance patterns regarding genuinely new and important legislation versus amending legislation. This also raises the question whether 'better' quality—more comprehensive, clear and consistent legislation—is better complied with, a question of particular relevance in light of the European Commission's Better Regulation Agenda and the recent 'Make it Work' Initiative aimed at harmonizing drafting provisions on compliance assurance in EU environmental law (Squintani 2016).

Second, enforcement and management approaches offer theoretical explanations of non-compliance in the EU. Rather than treating them as competing or alternative theories, our analysis shows that we can and should combine them. While this contribution has deliberately focused on EU-level developments, scholars should also apply these explanations at member state level. We need both more quantitative and qualitative research that systematically tests the (combined) explanatory power of enforcement and management and explores related causal mechanisms. Moreover, the literature has developed other theoretical explanations, for instance more sociological conceptions of compliance (Checkel 2001; Börzel et al. 2010; Börzel forthcoming). Public support for transferring legislative competences in a specific policy sector to the European level is highly relevant as EU citizens differ in degree of support for EU competencies over policy sectors; they are rather supportive of EU environmental policy-making but prefer their member states to retain control of employment and social affairs. However, if public acceptance matters for non-compliance, it might result in more rather than less non-compliance (Zhelyazkova et al. 2016). Environmental policy offers numerous research opportunities that can keep students of EU (environmental) policy busy for the next 25 years.

Acknowledgments

We thank the special issue editors as well as Viviane Gravey and Yaffa Epstein for their helpful comments on previous versions. We are particularly grateful to Andrea Lenschow for her excellent written feedback, Dimiter Toshkov for the data used in Figure 7 and to Lukas Blasius for his outstanding research assistance. We have also greatly benefited from discussion of earlier drafts at: the workshop 'Whither the environment in Europe?' ECPR Joint Sessions, Pisa, 24-28 April 2016; the workshop organized by Andreas Hofmann on 'The Future of Environmental Policy in the European Union' at the University of Gothenburg, 19–20 January 2017; and a dedicated panel at the European Union Studies Association (EUSA) Biannual Conference, Miami, FL, 3–5 May 2017.

Note

1. As our database includes only EU legislation starting from 1978, the first court referrals of these legal pieces are in 1981.

Disclosure statement

No potential conflict of interest was reported by the authors.

Funding

This work was supported by the German Research Foundation [BO 1831/1-1 and BO 1831/6-1].

ORCID

Aron Buzogány http://orcid.org/0000-0002-9867-3742

References

Andonova, L.B. and Tuta, I.A., 2014. Transnational networks and paths to EU environmental compliance: evidence from new member states. *Journal of Common Market Studies*, 52 (4), 775–793. doi:10.1111/jcms.12126

Angelov, M. and Cashman, L., 2015. Environmental inspections and environmental compliance assurance networks in the context of European Union environment policy. *In*: M. Faure, P. De Smedt, and A. Stas, eds. *Environmental enforcement networks: concepts, implementation and effectiveness*. Cheltenham: Edward Elgar, 350–376.

Angelova, M., Dannwolf, T., and König, T., 2012. How robust are compliance findings? A research synthesis. *Journal of European Public Policy*, 19 (8), 1269–1291. doi:10.1080/13501763.2012.705051

Bieber, R. and Maiani, F., 2014. Enhancing centralized enforcement of EU law: pandora's toolbox?'. *Common Market Law Review*, 51 (4), 1057–1092.

Bondarouk, E. and Mastenbroek, E., 2018. Reconsidering EU compliance: implementation performance in the field of environmental policy. *Environmental Policy and Governance*, 28 (1), 15–27. doi:10.1002/eet.v28.1

Börzel, T.A., 2001. Non-compliance in the European Union. Pathology or statistical artefact? *Journal of European Public Policy*, 8 (5), 803–824. doi:10.1080/13501760110083527

Börzel, T.A. and Risse, T., 2003. Conceptualising the domestic impact of Europe. *In*: K. Featherstone and C. Radaelli, eds. *The politics of Europeanisation*. Oxford: Oxford University Press, 55–78.

Börzel, T.A., 2003. *Environmental leaders and laggards in Europe. Why there is (not) a "Southern Problem"*. Hampshire: Ashgate.

Börzel, T.A., ed. 2009. *Coping with accession to the European Union. New modes of environmental governance*. Houndmills: Palgrave Macmillan.

Börzel, T.A., et al., 2010. Obstinate and inefficient. Why member states do not comply with European law. *Comparative Political Studies*, 43 (11), 1363–1390. doi:10.1177/0010414010376910

Börzel, T.A., forthcoming. *Non-compliance in the European Union*. Ithaca: Cornell University Press.

Börzel, T.A. and Buzogány, A., 2010. Environmental organisations and the Europeanisation of public policy in Central and Eastern Europe: the case of biodiversity governance. *Environmental Politics*, 19 (5), 708–735. doi:10.1080/09644016.2010.508302

Börzel, T.A. and Knoll, M., 2012. Quantifying non-compliance in the EU. A database on EU-infringement proceedings. Berlin Working Paper on European Integration No. 15, Freie Universität Berlin, Berlin Center for European Studies. Berlin: Berlin Center for European Integration.

Börzel, T.A. and Sedelmeier, U., 2017. Larger and more law abiding? The impact of enlargement on compliance in the European Union. *Journal of European Public Policy*, 24 (2), 197–217. doi:10.1080/13501763.2016.1265575

Braun, M., 2014. EU climate norms in East Central Europe. *Journal of Common Market Studies*, 52 (3), 445–460. doi:10.1111/jcms.12101

Bruszt, L. and Vukov, V., 2017. Making states for the single market: European integration and the reshaping of economic states in the Southern and Eastern peripheries of Europe. *West European Politics*, 40 (4), 663–687. doi:10.1080/01402382.2017.1281624

Buzogány, A., 2009. Hungary: the tricky path of building environmental governance. *In*: T. Börzel, ed. *Coping with accession to the European Union. New modes of environmental governance*. London: Palgrave, 123–147.

Buzogány, A., 2015. Building governance on fragile grounds: lessons from Romania. *Environment and Planning C: Government and Policy*, 33 (5), 901–918. doi:10.1177/0263774X15605897

Čavoški, A., 2015. A post-austerity European Commission: no role for environmental policy? *Environmental Politics*, 24 (3), 501–505. doi:10.1080/09644016.2015.1008216

Chayes, A. and Chayes, A.H., 1993. On compliance. *International Organization*, 47 (02), 175–205. doi:10.1017/S0020818300027910

Checkel, J.T., 2001. Why comply? Social learning and European identity change. *International Organization*, 55 (3), 553–588. doi:10.1162/00208180152507551

Collins, K. and Earnshaw, D., 1992. The implementation and enforcement of European Community legislation. *Environmental Politics*, 1 (4), 213–249. doi:10.1080/09644019208414052

Delreux, T. and Happaerts, S., 2016. *Environmental policy and politics in the European Union*. Houndmills: Palgrave Macmillan.

Dimitrova, A.L., 2010. The new member states in the EU in the aftermath of accession. Empty shells? *Journal of European Public Policy*, 17 (1), 137–148. doi:10.1080/13501760903464929

Dimitrova, A.L. and Buzogány, A., 2014. Post-accession policy making in Bulgaria and Romania: can non-state actors use EU rules to promote better governance. *Journal of Common Market Studies*, 52 (1), 139–156. doi:10.1111/jcms.12084

Downs, G.W., Rocke, D.M., and Barsoom, P.N., 1996. Is the good news about compliance also good news about cooperation? *International Organization*, 50 (3), 379–406. doi:10.1017/S0020818300033427

Duina, F.G., 1997. Explaining legal implementation in the European Union. *International Journal of the Sociology of Law*, 25 (2), 155–179. doi:10.1006/ijsl.1997.0039

Eder, K. and Kousis, M., 2001. *Environmental politics in Southern Europe: actors, institutions and discourses in a Europeanizing society*. Dordrecht: Kluwer.

European Commission, 2016. Communication from the European commission. EU law: Better results through better application. *C(2016)8600*. Brussels: European Commission.

Falkner, G., *et al.*, 2004. Non-compliance with EU directives in the member states: opposition through the backdoor? *West European Politics*, 27 (3), 452–473.

Falkner, G., Treib, O., and Holzleitner, E., 2008. *Compliance in the European Union. Living rights or dead letters?* Aldershot: Ashgate.

Fearon, J.D., 1998. Bargaining, enforcement, and international cooperation. *International Organization*, 52 (02), 269–305. doi:10.1162/002081898753162820

George, A.L. and Bennett, A., 2005. *Case studies and theory development in the social sciences*. Cambridge: MIT Press.

Gerdes, H., Davis, M., and Lukat, E., 2015. *The Implementation of the Natura 2000, Habitats Directive 92/43/ECC and Birds Directive 79/409/ECC*. Brussels: Committee of the Regions.

Haigh, N., 2015. *EU environmental policy: its journey to centre stage*. London: Routledge.

Hartlapp, M., 2007. On enforcement, management and persuasion: different logics of implementation policy in the EU and the ILO. *Journal of Common Market Studies*, 45 (3), 653–674. doi:10.1111/j.1468-5965.2007.00721.x

Hartlapp, M. and Falkner, G., 2009. Problems of operationalisation and data in EU compliance research. *European Union Politics*, 10 (2), 281–304. doi:10.1177/1465116509103370

Haverland, M., Steunenberg, B., and Warden, F.V., 2011. Sectors at different speeds: analysing transposition deficits in the European Union. *Journal of Common Market Studies*, 49 (2), 265–291. doi:10.1111/j.1468-5965.2010.02120.x

Hedemann-Robinson, M., 2016. Environmental inspections and the EU: securing an effective role for a supranational union legal framework. *Transnational Environmental Law*, 6 (1), 1–28.

Hille, P. and Knill, C., 2006. "It's the bureaucracy, stupid'. The implementation of the acquis communautaire in EU candidate countries, 1999–2003. *European Union Politics*, 7 (4), 531–552. doi:10.1177/1465116506069442

Hix, S., 2008. *What's wrong with the European Union and how to fix it*. Cambridge: Polity.

Hoffman, A., 2019. Left to interest groups? On the prospects for enforcing environmental law in the European Union. *Environmental Politics*, 28 (2). [this issue].

Holzinger, K., Knill, C., and Schäfer, A., 2006. Rethoric or reality? 'New governance' in eu environmental policy. *European Law Journal*, 12 (3), 403–420. doi:10.1111/j.1468-0386.2006.00323.x

Jehlicka, P. and Tickle, A., 2004. Environmental implications of Eastern Enlargement: the end of progressive EU environmental policy? *Environmental Politics*, 13, 77–93. doi:10.1080/09644010410001685146

Jordan, A., 1999. The implementation of EU environmental policy: a policy problem without a political solution? *Environment and Planning C*, 17 (1), 69–90. doi:10.1068/c170069

Jordan, A., 2001. National environmental ministries: managers or ciphers of European Union environmental policy? *Public Administration*, 79 (3), 643–663. doi:10.1111/padm.2001.79.issue-3

Jordan, A. and Adelle, C., 2012. *Environmental policy in the European Union: actors, institutions and processes*, 3rd edition. Abingdon: Taylor & Francis.

Knill, C., 1998. Implementing european policies: the impact of national administrative traditions. *Journal of Public Policy*, 18 (1), 1–28. doi:10.1017/S0143814X98000014

Knill, C. and Lenschow, A., 2000. Do new brooms really sweep cleaner? Implementation of new instruments in EU environmental policy. *In*: C. Knill and A. Lenschow, eds. *Implementing EU environmental policy: new directions and old problem*. Manchester: Manchester University Press, 251–282.

Knill, C., Steinebach, Y., and Fernández-i-Marín, X. 2018.Hypocrisy as a crisis response? Assessing changes in talk, decisions, and actions of the European Commission in EU environmental policy. *Public Administration*. Published online. doi:10.1111/padm.12542

König, T. and Luetgert, B., 2009. Troubles with transposition: explaining trends in member-state notification and the delayed transposition of EU directives. *British Journal of Political Science*, 39 (1), 163–194. doi:10.1017/S0007123408000380

König, T. and Mäder, L., 2014. The strategic nature of compliance: an empirical evaluation of law implementation in the central monitoring system of the European Union. *American Journal of Political Science*, 58 (1), 246-263. doi:10.1111/ajps.12038

Koops, C.E., 2011. EU compliance mechanisms: the interaction between the infringement procedures, IMS, SOLVIT and EU-PILOT. *Research Paper No. 2011-2042*. Amsterdam: Amsterdam Law School Legal Studies.

Koutalakis, C., 2004. Environmental compliance in Italy and Greece. The role of non-state actors. *Environmental Politics*, 13 (4), 754-774. doi:10.1080/0964401042000274359

Krämer, L., 2006. Statistics on environmental judgments by the EC Court of Justice. *Journal of Environmental Law*, 18 (3), 407-421. doi:10.1093/jel/eql019

Kronsell, A., 2002. Can small states influence EU norms?: insights from Sweden's participation in the field of environmental politics. *Scandinavian Studies*, 74 (3), 287-307.

La Spina, A. and Sciortino, G., 1993. Common agenda, Southern rules: european integration and environmental change in the Mediterranean States. *In*: J. D. Liefferink, P.D. Lowe, and A.P.J. Mol, eds. *European integration and environmental policy*. London, New York: Belhaven, 217-236.

Lauber, V., 1997. Austria: A latecomer which became a pioneer. *In*: M.S. Anderson and D. Liefferink, eds. *European environmental policy: the pioneers*. Manchester: Manchester University Press, 81-118.

Liefferink, D., et al., 2009. Leaders and laggards in environmental policy: A quantitative analysis of domestic policy outputs. *Journal of European Public Policy*, 16 (5), 677-700. doi:10.1080/13501760902983283

Mastenbroek, E., 2003. Surviving the deadline: the transposition of EU directives in the Netherlands. *European Union Politics*, 4 (4), 371-395. doi:10.1177/146511650344001

Mbaye, H.A.D., 2001. Why national states comply with supranational law. Exlaining impementation infringements in the European Union 1972-1993. *European Union Politics*, 2 (3), 259-281. doi:10.1177/1465116501002003001

Moravcsik, A., 2001. Despotism im Brussles. Misreading the European Union. *Foreign Affairs*, 80 (3), 114-122. doi:10.2307/20050155

Nugent, N., 2016. Enlargements and their impact on EU governance and decision-making. *Journal of Contemporary European Research*, 12 (1), 424-439.

Orru, K. and Rothstein, H., 2015. Not 'dead letters', just 'blind eyes': the Europeanisation of drinking water risk regulation in Estonia and Lithuania. *Environment and Planning A*, 47 (2), 356-372. doi:10.1068/a130295p

Panke, D., 2010. *The effectiveness of the European Court of Justice. Why reluctant states comply*. Manchester: Manchester University Press.

Peers, S., 2012. Sanctions for infringement of EU law after the Treaty of Lisbon. *European Public Law*, 18 (1), 33-64.

Pridham, G. and Cini, M., 1994. Enforcing environmental standards in the European Union: is there a Southern Problem? *In*: M. Faure, J. Vervaele, and A. Waele, eds. *Environmental standards in the EU in an interdisciplinary framework*. Antwerp: Maklu, 251-277.

Raustiala, K. and Slaughter, A., 2002. International law, international relations and compliance. W. Carlsnaes, T. Risse, and B.A. Simmons, eds. *The Handbook of International Relations*. London: Sage.

Schimmelfennig, F. and Sedelmeier, U., 2004. Governance by conditionality: EU rule transfer to the candidate countries of Central and Eastern Europe. *Journal of European Public Policy*, 11 (4), 661-679. doi:10.1080/1350176042000248089

Sissenich, B., 2007. *Building states without society: european Union enlargement and the transfer of EU social policy to Poland and Hungary*. Lanham: Lexington Books.

Skjærseth, J.B. and Wettestad, J., 2007. Is EU enlargement bad for environmental policy? Confronting gloomy expectations with evidence. *International Environmental Agreements: Politics, Law and Economics*, 7 (3), 263–280. doi:10.1007/s10784-007-9033-7

Smith, M., 2016. The visible, the invisible, the impenetrable: innovation or rebranding in centralised enforcement of EU law? *In*: S. Drake and M. Smith, eds. *New directions in the effective enforcement of EU law and policy*. Cheltenham: Edward Elgar, 45–76.

Sotirov, M., Lovric, M., and Winkel, G., 2015. Symbolic transformation of environmental governance: implementation of EU biodiversity policy in Bulgaria and Croatia between Europeanization and domestic politics. *Environment and Planning C: Government and Policy*, 33 (5), 986–1004. doi:10.1177/0263774X15605925

Spanou, C., 1998. Greece: administrative symbols and policy realities. *In*: K. Hanf and A.-I. Jansen, eds. *Governance and environment in Western Europe. Politics, policy and administration*. Essex: Longman, 110–130.

Squintani, L., 2016. Better regulation with make it work: an assessment of the make it work's drafting principles on compliance assurance. *Environmental Law Network International Review*, 2016 (1), 2–9.

Steinebach, Y. and Knill, C., 2017. Still an entrepreneur? The changing role of the European Commission in EU environmental policy-making. *Journal of European Public Policy*, 24 (3), 429–446. doi:10.1080/13501763.2016.1149207

Steunenberg, B. and Rhinard, M., 2010. The transposition of european law in EU member states: between process and politics. *European Political Science Review*, 2 (3), 495–520. doi:10.1017/S1755773910000196

Tallberg, J., 2002. Paths to compliance: enforcement, management, and the European Union. *International Organization*, 56 (3), 609–643. doi:10.1162/002081802760199908

Toshkov, D., 2010. Taking stock: A review of quantitative studies of transposition and implementation of EU law. *eif Working Paper 01/2010*.

Tosun, J., 2012. Environmental monitoring and enforcement in Europe: A review of empirical research. *Environmental Policy and Governance*, 22 (6), 437–448. doi:10.1002/eet.v22.6

Treib, O., 2014. Implementing and complying with EU governance outputs. *Living Rev. Euro. Governance*, 9 (1). Available from: http://www.europeangovernance-livingreviews.org/Articles/lreg-2014-1/#tabs-3.

Zhelyazkova, A., Kaya, C., and Schrama, R., 2016. Decoupling practical and legal compliance: analysis of member states' implementation of EU policy. *European Journal of Political Research*, 55 (4), 827–846. doi:10.1111/1475-6765.12154

Zhelyazkova, A., Kaya, C., and Schrama, R., 2017. Notified and substantive compliance with EU law in enlarged Europe: evidence from four policy areas. *Journal of European Public Policy*, 24 (2), 216–238. doi:10.1080/13501763.2016.1264084

Zito, A.R., 2000. *Creating environmental policy in the European Union*. London: Macmillan.

Zito, A.R., Burns, C., and Lenschow, A., 2019. Is the trajectory of European Union environmental policy less certain? *Environmental Politics*, 28 (2). [this issue].

Left to interest groups? On the prospects for enforcing environmental law in the European Union

Andreas Hofmann

ABSTRACT
There have been important changes in the enforcement of European Union environmental law over the last 25 years. Environmental law has traditionally been reliant on the European Commission, but the Commission has started to withdraw from enforcement. Instead, it is undertaking efforts to 'outsource' enforcement to environmental non-governmental organisations (NGOs) by systematically promoting access for such groups to national courts. While the Commission has indicated that it sees centralised and private enforcement as substitutes, the advantages and drawbacks of each mechanism are evaluated and it is concluded that both mechanisms have an important role to play. In particular, the private enforcement of EU environmental law is dependent on national opportunity structures that are unlikely to ever be fully liberalised and harmonised by EU procedural law. Private enforcement is therefore not a panacea for compliance problems, and the growing absence of a central enforcing authority is a cause for concern.

Introduction

Writing about European Union (EU) environmental law in *Environmental Politics* in 1992, Ken Collins and David Earnshaw (1992, p. 214) observed that 'legislation will not be worth the paper it is printed on if policies break down or obligations are not fulfilled at the implementation stage'. While the problem of assuring central policies' practical effect 'on the ground' is timeless (e.g. Pressman and Wildavsky 1974), Collins, then chairman of the European Parliament's committee on the environment, and his co-author stressed that attention to implementation of EU environmental law was in 1992 a relatively recent phenomenon. The national implementation of EU law is far from automatic. In the case of EU directives (the major legal instrument in EU environmental legislation), it consists of both a formal transposition into national law and an adaptation of national practices that lead to tangible results (Collins and Earnshaw 1992, pp. 215–216; Bondarouk and Mastenbroek 2017; Thomann and Sager 2017). Many different national actors are involved in this process, from legislators to street-level bureaucrats. Results are often not in line with a policy's original intention, for many reasons, including: lack of national administrative capacities, lack of salience, a 'misfit' between the EU policy and national policy traditions, or powerful national interests opposing the policy (Borrass et al. 2015; Versluis 2007; Treib 2014). All EU policies are subject at times to serious implementation deficits (Falkner et al. 2004; Börzel et al. 2010). Any discussion of policy implementation is therefore closely connected to enforcement as a means of remedying apparent implementation failures.

Here, I highlight important changes in how EU environmental law has been enforced over the last 25 years. My central argument is that the Commission, long the central actor in the enforcement of EU environmental law, is increasingly withdrawing from this role and instead attempts to 'outsource' enforcement to private actors, in particular to environmental non-governmental organisations (NGOs). While these groups have traditionally faced significant obstacles to pursuing environmental litigation in the courts of many member states, EU legislation implementing the Aarhus Convention has recently liberalised national procedural rules. The Commission has supported this development through both legislative initiatives and judicial interventions. I present these efforts in detail and discuss whether the increased opportunities for private enforcement can compensate for the reduced engagement of the Commission.

I structure the contribution as follows. I first juxtapose two mechanisms for the judicial enforcement of EU environmental law, highlighting the traditional centrality of the Commission in this field and discussing the deficits of Commission enforcement. Second, I examine Commission enforcement priorities over time and present data indicating the extent of the Commission's retrenchment from centralised enforcement. Third, I trace the expanding opportunity structure for de-centralised enforcement of environmental law, particularly increased access to national courts for environmental NGOs in the wake of the Aarhus Convention. Finally, I conclude by discussing the deficits of this process, highlighting its uneven effects across member states, and the resulting need for a continuing presence of the Commission.

Centralised and de-centralised enforcement

EU treaties allow for two mechanisms by which national compliance with EU law can be judicially enforced: a centralised mechanism in which the Commission initiates infringement proceedings against non-complying member states, and a de-centralised mechanism in which citizens, groups and companies initiate legal proceedings against non-complying national authorities before national courts. National courts have the possibility to refer questions about the proper interpretation of EU law to the EU Court of Justice (CJEU), and such 'preliminary references' constitute a significant part of the CJEU's caseload. National courts also have the possibility to enforce EU law directly. The de-centralised (private) enforcement procedure is what makes the EU legal order unique among international legal orders; its active use by citizens and companies has contributed significantly to EU law's impact 'on the ground'.

EU environmental law has, however, always been an outlier in this regard. Compared to other policy areas, its enforcement is much more reliant on the centralised mechanism of Commission infringement proceedings. Litigation initiated by the Commission typically makes up between 15% and 40% of the CJEU's overall caseload in any given year, the remainder primarily consisting of preliminary references from national courts in cases initiated by private parties (Figure 1). These figures look very different for the environmental sector. With few exceptions, between 1992 and 2010 over 60% of all environmental cases that reached the CJEU were brought by the Commission as infringement cases. Between 1999 and 2006, the share of infringement procedures was over 80%. Since 2010, the percentage has fallen steeply but still remains about 20 percentage points above the average.

Unlike many other policy areas, EU environmental law does not vest citizens with individual rights—such as a right to clean air or water—that

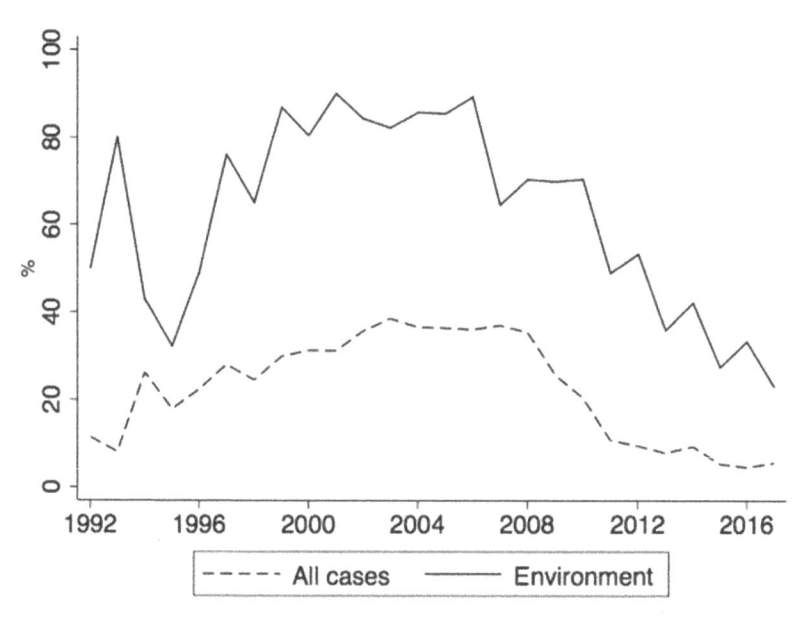

Figure 1. Share of Commission-initiated cases before the CJEU.
Note: Data based on annual reports of the CJEU 1992–2017 and the curia.europa.eu database.

could activate access to national courts. Instead, EU environmental law has been drafted as protecting a public good (for a thorough discussion of the lack of rights frames in EU environmental law, see Hilson 2017). Litigation in the public interest has traditionally been absent in most European countries. Barred from access to national courts, citizens and environmental NGOs often had little alternative but to complain to the Commission, and contacting the Commission is still part of the standard procedure of many environmental NGOs. The comparatively high number of complaints the Commission receives about breaches of EU environmental law reflects this situation. According to the Commission's 'annual reports on monitoring the implementation of EU law' (hereafter 'annual monitoring reports'), the environmental sector regularly ranks among the four policy areas generating most complaints.

Deficits of centralised enforcement

The debate about the obstacles to private litigation in environmental matters is longstanding. Collins and Earnshaw (1992, p. 244) highlighted that restrictive access to courts made private enforcement of EU environmental law 'fraught with difficulties' and suggested that a 'measure enabling environmental non-governmental organisations and private individuals to bring cases for practical infringements of EC environment legislation in national courts would be an important measure which the Commission might usefully consider'. However, what are the problems with reliance on the Commission that private enforcement could potentially redress?

Centralised Commission enforcement has numerous advantages. First and foremost, the Commission is comparatively well funded, experienced and well staffed to litigate environmental cases. It wins almost all the cases it brings to the CJEU (Börzel et al. 2012, p. 456), a success rate that environmental NGOs can only dream of (Darpö 2012, p. 11). The possibility for financial penalties in cases of protracted member state non-compliance has added to the Commission's ability to effectively enforce EU law, and the environmental sector is one of the main beneficiaries. Of the 33 cases to date in which the CJEU has decided on financial penalties, 14 have concerned environmental law.[1] Infringement proceedings are also salient, not least since the Commission started issuing press releases with all its infringement decisions. Aggregate statistics about procedures initiated and cases litigated against individual member states are readily available, making for relatively easy 'naming and shaming'. In comparison, data on the use of EU environmental law in national courts are scarce and aggregate statistics for actual incidences of private enforcement are unavailable except for the relatively few cases that get referred to the CJEU for a preliminary ruling.

Nevertheless, centralised enforcement also has numerous important drawbacks. First, the procedure is slow. The Commission states the average duration of an infringement procedure as 26 months, but cases that it refers to the CJEU can last over 50 months overall (European Commission 2007, p. 4). National court cases do not necessarily take less time—there is considerable variation across member states—but the duration of centralised enforcement is particularly problematic since the opening of an infringement procedure has no suspensive effect. Once the Commission has referred a case to the CJEU it can ask for the immediate suspension of the activity in question ('interim relief'), but it uses this option rarely, and at this stage it may already be too late (Hedemann-Robinson 2010). Potentially harmful activity will have continued for many months, large projects will have been built and protected species already hunted. As the Commission itself notes, private enforcement can achieve better and faster remedies: 'Only a national tribunal can apply remedies like injunctions to the administrations, cancellation of national decisions, damages, *etc.*' (European Commission 2007, p. 8).

Another limitation lies in the Commission's limited ability to monitor the national implementation of EU environmental law. The Commission relies heavily on citizens and NGOs to flag possible cases of non-compliance (European Commission 2016a, p. 2). The reality of centralised enforcement is therefore closer to a constant fire alarm rather than the police patrol it is often portrayed as (Tallberg 2002). Moreover, the Commission does not pursue all infringements it is alerted to but follows its own priorities in selecting cases to pursue; these priorities vary over time. The Commission has wide discretion about using the infringement procedure (Eliantonio 2016, pp. 179–180). It is under no obligation to pursue an infringement, even where a member state's breach of obligations is obvious, and retains the right to terminate the procedure at any point regardless of the member state's compliance with its demands (Craig and de Búrca 2011, p. 415). The current CJEU president Koen Lenaerts described the procedure as 'a political tool at the Commission's disposal' (Lenaerts and Gutiérrez-Fons 2011, p. 4).

The private enforcement of EU environmental law, on the other hand, is independent of the Commission's priorities. From the affected citizens' perspective, directly initiating legal action in a national court can seem more immediately effective than contacting the Commission and waiting for it to act. The ability of individual citizens or NGOs to bring cases to national courts, however, depends on the 'opportunity structure' (Kitschelt 1986; Hilson 2002) offered by national legal systems for bringing legal action. This not only covers rules on legal standing but also aspects such as the cost and duration of legal proceedings and the types of remedies available (Conant *et al.* 2017). National legal systems rarely offer favourable conditions on all these aspects. In order to allow for effective de-centralised enforcement of EU environmental law, EU legislation would need to both harmonise and liberalise existing national opportunity structures. The persistence of great variation in the conditions for private enforcement of EU environmental law constitutes the central drawback of this mechanism. As shown below, legal orders are slow to change.

The following sections address both mechanisms in more detail. The next section studies the extent to which the enforcement priorities of the Commission pose a problem for effective enforcement of EU environmental law. I then trace the introduction, promoted by the Commission, of procedural rights aimed at enhancing access to national courts for citizens and environmental NGOs, and discuss the degree to which this effort succeeded in removing obstacles to de-centralised enforcement of EU environmental law.

Commission enforcement priorities

EU environmental law's traditional dependence on the Commission means that its enforcement is closely bound to the Commission's priorities in pursuing non-compliance. This may be problematic, as recent events illustrate (Čavoški 2015). When Jean-Claude Juncker in 2014 presented the priorities for his new Commission as an agenda for jobs and growth, environmental groups were quick to point out that, with the exception of a reference to climate change, environmental protection was largely absent (European Environmental Bureau 2015, p. 6; WWF 2015b, p. 13). This concern strengthened when Juncker merged the Commission environment portfolio with that for maritime affairs and fisheries. The Commission reinforced this worry when it announced that it would subject the EU's birds and habitats directives (European Commission 2014), two of the central pieces of EU environmental legislation, to its 'Regulatory Fitness and Performance Programme' (REFIT), which aims at 'removing red tape and lowering costs' (European Commission 2015a). The WWF felt compelled to highlight that 'this process is happening in a context that is clearly hostile to nature conservation, as President's Juncker rhetoric on "business-friendly" laws and cutting "green tape"' illustrated (WWF 2015a).

The Juncker Commission by contrast has emphasised the compatibility of environmental law with an agenda for growth. In its latest annual monitoring report, the Commission subsumed enforcement of environmental law under its priority of 'a new boost for jobs, growth and investment' and highlighted that EU environmental law 'helps to ensure a level playing field for all Member States and economic operators that need to meet the environmental requirements. Strict enforcement also stimulates the market to find innovative ways to increase resource efficiency and reduce import dependency. Such innovation can give EU companies a competitive edge and create jobs' (European Commission 2016b, p. 5).

While this is unlikely to significantly calm the nerves of environmental groups, recent data give no indication that the re-adjustment of the Commission's priorities has particularly de-emphasised the enforcement of environmental law. The environment has traditionally constituted one of the Commission's central enforcement priorities and apparently continues to do so. According to the Commission's annual monitoring reports, between 1998 and 2015 the environment has, with very few exceptions, been the policy area with the highest share of open infringement proceedings. On average, 23% of open cases consistently pertained to environmental matters during this time. The pattern for cases of persistent conflict—those cases the Commission has referred to the CJEU—also yields no evidence that the Commission is de-prioritising the environment relative to other policy areas. The first 2 years of the Juncker Commission replicate a pattern where environmental cases constitute about 20% of all cases that the Commission litigates.[2]

EU environmental law's vulnerability to the Commission's changing enforcement priorities comes in a different guise. Since 2010, there has been a steep fall in all indicators of Commission enforcement activity, for environmental cases as for all other policy fields. From 2011 to 2015 the number of open cases in environmental matters was at its lowest since the late 1990s (Figure 2); the number of cases the Commission referred to the CJEU was as low as in the first half of the 1990s (Figure 3) when the EU had considerably fewer members than

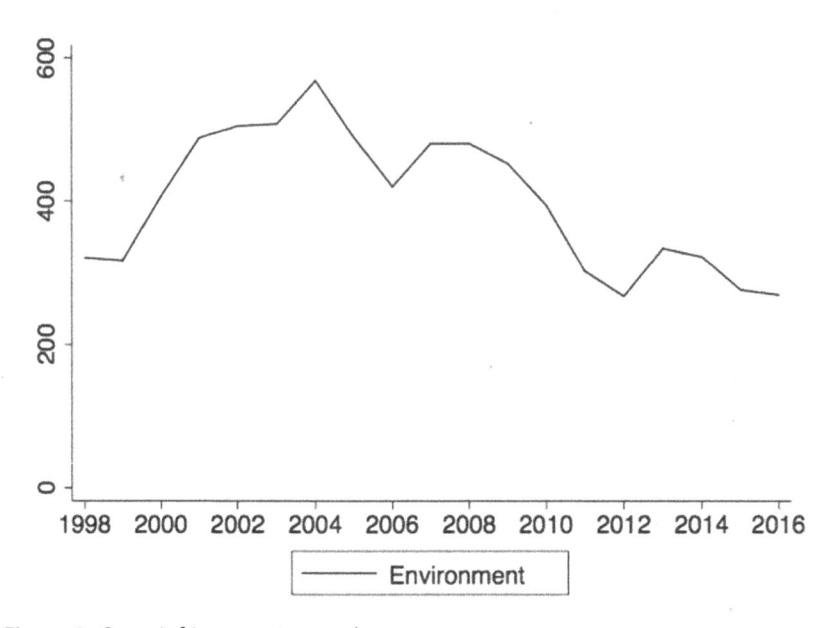

Figure 2. Open infringement procedures.
Note: Data based on the Commission's annual monitoring reports 1998–2016.

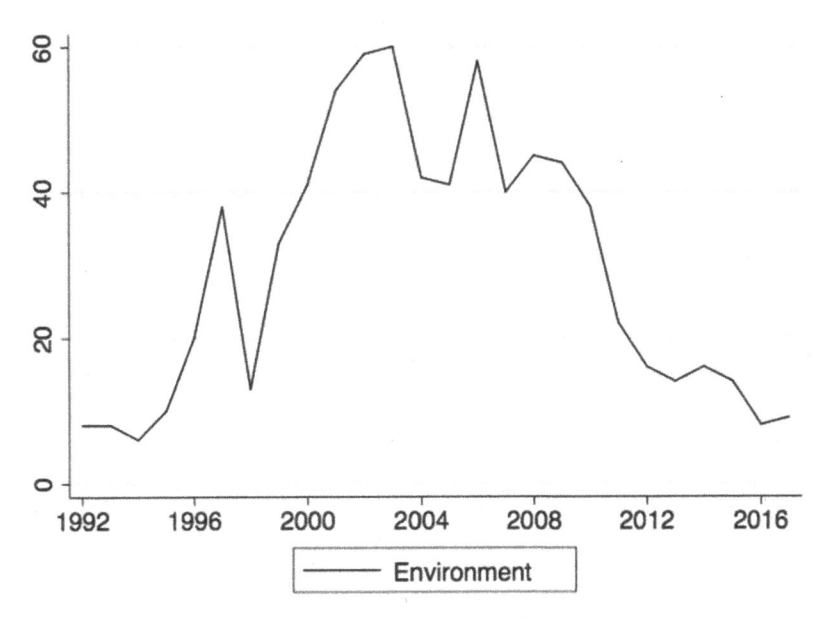

Figure 3. Number of infringement cases referred to the CJEU.
Note: Data based on annual reports of the CJEU 1992–2017 and the curia.europa.eu database.

today. This general decline holds also when looking only at cases of substantive conflict (excluding cases the Commission routinely initiates when member states fail to notify a transposition measure for a directive). Between 2003 and 2009, the Commission referred an average of 24 environmental cases to the court concerning substantive non-compliance with environmental law. This average dropped to 13 cases between 2010 and 2016.[3]

There are several explanations for the Commission's withdrawal from centralised enforcement. The timing of the decline coincides with the introduction of 'EU Pilot', the Commission's mechanism for the informal handling of complaints before initiating a formal infringement procedure. EU Pilot has faced important criticism for delaying Commission action (Smith 2016); a former head of the legal unit of the Commission's Directorate-General Environment even accused EU Pilot of having been designed to deter complaints (Krämer 2014, p. 251). However, the Commission itself is reluctant to identify EU Pilot as the cause of the decrease in enforcement activity (European Commission 2011, p. 6). Moreover, data from the Commission's annual monitoring reports on the use of EU Pilot show that newly opened EU Pilot cases also declined significantly after 2013. This trend could indicate that member states have improved their compliance record (Börzel and Buzogany 2019). However, recent implementation reports for EU environmental law, including the Commission's own, outline substantial compliance problems—even with older pieces of environmental legislation, which indicates that even a decrease in new environmental legislation cannot wholly account for the decrease in enforcement measures (IMPEL 2015, European Commission 2017c). The development is however consistent with an increasingly restrained policy activism that has been identified since the second Barroso Commission (Kassim et al. 2017, p. 666). It also coincides with EU environmental policy's entry into a new stage, characterised by rapid contextual change and increasing instability since the start of the economic and financial crisis (Zito et al. 2019).

The Commission itself offers yet another different explanation. It explicitly links the reduction of its own efforts to an increase in private cases that reach the CJEU: 'The overall decrease of the number of infringement procedures can be put in relation to the important increase of preliminary rulings under Article 267 TFEU since 2010' (European Commission 2015b, p. 16). This is true both for the overall amount of cases and those in the environmental sector. The next section outlines the legal developments that contributed to this situation; then I highlight the problem with the Commission's apparent belief that private enforcement can substitute for its own centralised efforts.

Opportunities for de-centralised enforcement

Where early EU documents touched upon issues of member state compliance with EU environmental measures, such as its first 'action programme' on the environment in 1973, the emphasis was on the Commission's role in monitoring and enforcing (Council 1973, p. 30). Private enforcement of environmental law only became a topic in Commission documents around the time of the United Nations Conference on Environment and Development in Rio de Janeiro in 1992. The resulting 'Rio Declaration', agreed to by all EU member states, stated in principle 10 that 'effective access to judicial and administrative proceedings [for citizens], including redress and remedy, shall be provided'. In the environmental law section of its 1992 annual monitoring report, the Commission stressed that 'the rise in the number of complaints from Europe's citizens is evidence of [...] their limited access to the courts' (European Commission 1993, p. 41). The Fifth Environmental Action Programme stated in 1993 that 'individuals and public interest groups should have practicable access to the courts in order to ensure that their legitimate interests are protected and that prescribed environmental measures are effectively enforced' (Council 1993, p. 82). From this point on, appeals to facilitate de-centralised enforcement, or increase 'access to justice', became routine. In its 1994 annual monitoring report, the Commission stated that it was considering 'an instrument to facilitate public involvement in the application of Community environmental law via direct access to justice' (European Commission 1995, p. 47). In 1996, the Commission devoted a whole section to access to justice in its communication on 'implementing Community environmental law', in which it acknowledged both the limitations of the centralised approach and the obstacles that private enforcement of environmental law faced in national legal systems. Not only were national procedural rules unfavourable, the environmental sector also suffered from a 'frequent lack of a private interest as an enforcement driving force' (European Commission 1996, p. 12). Therefore, it would be 'necessary to look wider than individuals directly affected and include representative organisations seeking to protect the environment', and to give environmental NGOs standing in national courts against national authorities, much in the mould of similar rules concerning consumer protection (European Commission 1996, p. 13).

The Aarhus Convention

Around the time of the 1996 Commission communication, negotiations had started in the framework of the United Nations Economic Commission for Europe on a 'convention on access to information, public participation in decision-making and access to justice in environmental matters'. The product of these negotiations, the Aarhus Convention, was signed in 1998 and subsequently ratified by the EU and all EU member states. It confers rights to inclusion in decision-making and private enforcement on the 'public' in the sense of natural and legal persons, their associations, organisations and groups, including NGOs promoting environmental protection (article 2.4 and article 2.5, Aarhus Convention). As an international treaty, the provisions of the Convention did not automatically create new rights; both the EU and member states had to implement the provisions.

The Convention provides for access to justice in several respects. In a first step, rights to information and participation in decision-making procedures are enforceable in court. In a second step, the Convention provides for substantive and procedural complaints against national permitting processes and environmental impact assessments concerning large construction projects (residential developments, roads, power lines, power plants). The Commission in 2000 proposed two directives to implement these two steps, which passed the EU legislative process with relatively little conflict (resulting in Directive 2003/4 on public access to environmental information and Directive 2003/35 on public participation in planning processes, respectively). In a third step, the Convention mandates that members of the public, including interest groups, 'have access to administrative or judicial procedures to challenge acts and omissions by private persons and public authorities' that violate environmental laws (article 9.3, Aarhus Convention). This article, if implemented, would allow interest groups to go to court against third parties, including private enterprises, that infringe on any environmental law. It is therefore perhaps not surprising that this article proved the most contentious in the process of implementing the Aarhus Convention. The Commission's 2003 proposal for a directive providing wide access to justice for interest groups in environmental matters met with opposition in both the European Parliament and the Council, despite the fact that it excluded the possibility to go to court against private entities (Poncelet 2012, p. 291). The proposal did however state that 'entities active in the field of environmental protection [...] should have access to environmental proceedings in order to challenge the procedural and substantive legality of administrative acts and omissions which contravene environmental law' (European Commission 2003, recital 9). This would allow interest groups to go to court against public bodies where they fail to act against environmental pollution or destruction. Since the EU legislative institutions could not agree on the proposal for a number of years (member state governments, in particular, expressed concerns for the integrity of their judicial systems), the Commission withdrew the proposal in May 2014. Despite repeated calls by environmental NGOs (e.g. European Environmental Bureau 2015, p. 33), the Commission is not planning to submit a new legislative proposal, but has issued an 'interpretative communication' on existing legislation and case-law (European Commission 2017a). The Aarhus Convention's most far-reaching provision will therefore not be transposed into EU legislation.

Procedural autonomy and the heterogeneity of access to justice

While the Commission has long championed the de-centralised enforcement of EU environmental law, the lack of a broad EU competence to regulate national procedural law, i.e. the 'principle of (national) procedural autonomy', limits concrete efforts. The CJEU has repeatedly emphasised that national procedures must constitute an effective means of enforcing EU rights, and that it must not be more difficult to enforce an EU right than a national right (Craig and de Búrca 2011, pp. 218–220), but national legal systems provide very different opportunity structures to bring environmental cases to court, even when enforcing national environmental law. With growing attention to access to justice in environmental matters in the wake of the Aarhus Convention, several cross-national comparative studies assess the state of the procedural diversity within the EU (Prieur 1998; IMPEL 2000). The central questions along which legal systems were found to diverge were: who has standing to appeal, what types of decisions can be appealed, what arguments are admissible, how long do procedures take, what remedies are available, and what are the costs of the proceedings? (Prieur 1998, p. 24)

Before the ratification of the Aarhus Convention, several countries (Austria, Denmark, Germany and Sweden) did not grant standing to environmental NGOs to challenge administrative decisions of any kind (although some German Länder had started to do so under their environmental laws, Prieur 1998, pp. 11–12). The Nordic countries relied primarily on a tradition of appeal to non-judicial bodies, whereas Austria and Germany (and, to some degree, Italy) allowed for judicial review only under restrictive rules of standing and restricted the admissible grounds of appeal to arguments that had previously been used in administrative procedures. Other legal systems allowed for wide access. Ireland and the United Kingdom had liberal standing rules, but costly procedures, 'loser pays' rules (where the losing party has to pay the legal expenses of the winning party) and potential liability for economic losses due to delay, limited litigation. The studies also named costs as significant obstacles in Italy and Germany (Prieur 1998). France, Greece, Spain and Portugal offered generous standing (Portugal generally allowed for public interest litigation), but courts in these countries (and Italy) were comparatively slow to process cases.

This diversity meant that the Aarhus Convention would have very different effects across national legal systems and that those effects would likely be strongest where national procedural rules were most restrictive. However, the Convention's stipulation that judicial procedures should provide 'adequate and effective remedies, including injunctive relief as appropriate, and be fair, equitable, timely and not prohibitively expensive' (article 9.4, Aarhus Convention) required all national legal systems to adapt to some degree. Since the Convention is about procedural guarantees, EU legislation implementing the Convention entailed harmonisation of national procedural laws concerning the environment. This legislation can itself be enforced through the EU's regular mechanisms. Importantly, and contrary to substantive environmental law, this procedural environmental law conferred on citizens and NGOs distinct rights that would be enforceable in national courts. Some of these procedural rights are contained in substantive legislation such as the Environmental Impact Directive and the Environmental Liability directive, thereby introducing the EU's strong individual rights regime 'through the back door' to an area where it had traditionally been absent. Eliantonio and Muir (2015) have called this practice an 'incidental proceduralisation' of EU environmental law. Both the Commission and NGOs themselves have used this new EU procedural rights regime against restrictive national procedural rules in order to facilitate the de-centralised enforcement of substantive EU environmental law (Oliver 2013, pp. 1446–1455).

Litigating procedural rights

As could be expected, efforts to expand procedural rights have occurred in countries with the most restrictive rules, particularly concerning standing, admissible arguments and costs. A Swedish NGO, Djurgården-Lilla Värtans Miljöskyddsförening, involved in a case concerning construction permits for power lines, challenged a Swedish rule stipulating that such groups would only have access to courts if they had carried out activities in Sweden for at least 3 years and had more than 2000 members.[4] Sweden was one of only a few EU countries to apply such a numerical criterion for interest group standing; the government itself conceded that only two environmental groups in Sweden at the time met the criteria (Oliver 2013, p. 1450). On appeal, the Swedish Supreme Court referred the question to the CJEU, which in October 2009 decided that 'national rules must not be liable to nullify Community provisions which provide that [...] environmental protection associations are to be entitled to bring actions before the competent courts' (CJEU 2009, para 45). In reaction, the Swedish government adjusted the applicable law in July 2010 and reduced the membership requirement to 100 (Swedish Code of Statutes 2010).

In 2008, the German environmental NGO, BUND, involved in a dispute over the construction of a coal-fired power plant, contested a German principle of administrative law ('*Schutznormtheorie*') that administrative decisions can only be challenged if they infringe on legislative provisions conferring individual rights (as opposed to public interests, Oliver 2013, p. 1452). The responsible German administrative court expressed doubts whether this principle was in conformity with EU law and referred the question to the CJEU, which in May 2011 agreed that NGOs should be able to challenge industrial permits based on public interests (CJEU 2011a, para. 50). In response, the German government adapted parts of the Environmental Appeals Act in November 2012 to grant wider standing for interest groups.

Private litigants also challenged the comparatively high costs of court cases in the UK. Since the losing party in legal proceedings was potentially liable for the total legal costs of the winning party, the costs of a case, particularly if it went to appeal, could be very high. In 2011, the UK Supreme Court referred to the CJEU the question whether this rule conflicted with the Aarhus Convention's requirement that judicial review should not be 'prohibitively expensive'. The case had started as a conflict about a permit for a cement factory and went through several stages of appeal. All decisions went against the private citizens who had challenged the permit. After the last appeal at the highest level of the UK court system failed, the winning parties (the permit-issuing authorities) submitted a bill amounting to about 88,000 GBP, for which the main litigant would potentially be liable. The Supreme Court asked the CJEU whether this would run counter to the Aarhus Convention, and the CJEU replied in 2013 that costs must not act as a deterrent to potential litigants in environmental matters (CJEU 2013, para. 35). A cost decision should moreover take account of the litigant's personal circumstances: costs should neither be objectively nor subjectively unreasonable (De Baere and Nowak 2016, p. 1734).

Around the same time, the Commission had also taken the UK government to the CJEU over the cost of legal procedures (CJEU 2014). Its case raised similar arguments, with a particular emphasis on the predictability of a legal procedure's final costs for the litigant before she engages in it (De

Baere and Nowak 2016, p. 1735). While both cases were still pending, the UK enacted changes in its procedural rules that capped the potential costs of environmental cases at 5,000 GBP for individuals and 10,000 GBP for organisations.

Since the signing of the Aarhus Convention, the Commission has supported the newly established procedural rights of private litigants in environmental matters in a number of infringement proceedings against member states. It had previously taken the Irish government to court about the high costs of Irish legal procedures, with the result that Ireland no longer applies the 'loser pays' rule in environmental cases (Ryall 2013, p. 4). It also initiated infringement procedures against restrictive rules on standing, such as the German rule of 'material preclusion', which German environmental NGOs identified as particularly onerous. According to this rule, legal challenges to administrative decisions could only rely on arguments that had already been raised in detail during administrative consultation (Bunge 2016, p. 19). Given tight consultation deadlines, the practical effect of this rule was that environmental interest groups needed to invest considerable resources at a very early stage of each permitting procedure in order not to have to forgo important arguments in a possible legal challenge. In its judgment of October 2015, the CJEU agreed with the Commission that a certain reduction in administrative efficiency would have to be accepted in order 'not only to ensure that the litigant has the broadest possible access to review by the courts but also to ensure that that review covers both the substantive and procedural legality of the contested decision in its entirety' (CJEU 2015, para. 79–80).

These cases demonstrate that NGOs, individual litigants and the Commission have been successful in slowly chipping away restrictive national procedural rules. Nonetheless, more recent comparative overviews of national procedural law show that these efforts have only limited effect across member states (Milieu 2007; Darpö 2013). CJEU case law in environmental matters appears to have little impact outside the national context from which the case emerged. Exceptions are cases where the national rule in question has a close equivalent in other legal systems, such as the norm of material preclusion in Austria (whose government intervened in the case against Germany). Relatively few cases have considerable implications outside their national contexts.

One such case concerned the question whether interest groups have a general right of access to justice in environmental matters, despite the fact that this provision of the Aarhus Convention had not been transposed into EU legislation. The underlying conflict was a challenge by a Slovak environmental NGO (VLK Lesoochranárske zoskupenie) against a decision by public authorities to issue permits to hunt brown bears, a species protected under the EU's Habitats Directive. Slovak law did not grant standing for NGOs in such cases. The Slovak Supreme Court referred the question to the CJEU, which stated in its March 2011 judgment that 'it is for the referring court to interpret, to the fullest extent possible, the procedural rules [...] so as to enable an environmental protection organisation [...] to challenge before a court a decision [...] liable to be contrary to EU environmental law' (CJEU 2011b, para. 50–51). The Slovak Supreme Court followed suit and granted the group standing to challenge the hunting licenses (Brakeland 2014, p. 15).

This 'Slovak bears' case has had considerable impact in countries with restrictive traditions of legal standing, such as Germany or Sweden (but so far not in Austria). National courts in these countries have started applying the CJEU's reasoning to set aside unfavourable national procedural rules and allow interest groups to challenge all sorts of administrative decisions relating to environmental matters. This has been the case regarding German rules on air pollution, particularly limits on nitrogen dioxide. EU rules on air pollution require that, where limits for nitrogen oxide are exceeded, national authorities must draw up a short-term action plan indicating how emission levels can be brought in line with existing limits (e.g. by way of a congestion tax or restrictions on diesel cars). In previous case law, the CJEU had held that individuals ('persons directly concerned') could bring a legal challenge if such an action plan was missing, setting aside German rules to the contrary (CJEU 2008, para. 39), but there was no equivalent right of standing for interest groups. Throughout 2011 and 2012, a German environmental NGO (Deutsche Umwelthilfe) tested this limitation by bringing a number of cases itself. The responsible lower administrative court relied explicitly on the CJEU's ruling in the 'Slovak bears' case when it granted the NGO standing and ordered the state government to revise its action plan, a ruling that the Federal Administrative Court upheld on appeal (Bundesverwaltungsgericht 2013). Deutsche Umwelthilfe has since, with the help of ClientEarth (Goodman and Connelly 2018), challenged multiple local action plans relating to nitrogen dioxide pollution (Clean Air 2017).

Administrative courts in Sweden have also taken up the CJEU's judgment in the 'Slovak bears' case. The first such case concerned a 2011 hunting permit that the Swedish Environmental Protection Agency had issued for the culling of a wolf. The Swedish Society for Nature Conservation (SSNC), one of Sweden's largest environmental NGOs, appealed this decision. Eventually, the Supreme Administrative Court granted standing, specifically referring to the 'Slovak bears' case in setting aside restrictive Swedish procedural rules (Högsta Förvaltningsdomstolen 2012, see also Epstein and Darpö 2013, pp. 258–260). The SSNC has since challenged numerous permits for the hunting of protected large carnivores.

Swedish administrative courts have moreover started to expand the same reasoning to cases not otherwise related to EU environmental law, such as forestry. The first such case concerned the planned clear-cutting of a privately owned forest that was held to include valuable habitat. SSNC appealed the initial permit decision to an administrative court, which granted SSNC standing with reference to the Aarhus Convention, and voided the permit because of the area's valuable status (Förvaltningsrätten i Luleå 2011), a decision that the Supreme Administrative Court upheld (Högsta Förvaltningsdomstolen 2014). This decision suggests that environmental organisations now face few obstacles to challenge any administrative decision in Sweden that may have environmental consequences (Darpö 2014, p. 5).

Conclusions: can NGOs replace the commission?

As these developments show, the Aarhus Convention and its transposition into EU law have improved opportunities for the de-centralised enforcement of EU environmental law. Procedural rights have enabled NGOs to expand the legal opportunity structures to enforce substantive environmental law. Litigation by environmental NGOs is now possible where previously it was not. Rights of standing, procedural rules and cost burdens have generally improved. Individual examples, such as the numerous air pollution cases, show that increased access to courts can also have substantial impact on member state compliance with environmental obligations. Nevertheless, while de-centralised enforcement has certainly broadened the options for citizens and environmental groups, it is by no means a panacea for environmental law's persistent implementation failures. As demonstrated above, distinct differences in opportunities for the private enforcement of environmental law remain, despite both legislative and judicial efforts to liberalise and harmonise procedural rules (Eliantonio 2015). Legal traditions are sticky. Procedures remain comparatively expensive in Anglo-Saxon countries, and German-speaking countries are slow to adapt their systems of administrative law, which are traditionally hostile to public interest litigation. Other member states, such as the Netherlands, have enacted restrictive reforms after ratification of the Aarhus Convention. This diversity has led one prominent expert on EU environmental law to summarise national opportunity structures for private enforcement as 'diverging, random and inconsistent' (Darpö 2013, p. 11).

Perhaps the clearest examples that the efforts to liberalise procedural rules cannot relieve the Commission of its enforcement role are found in some Mediterranean EU member states. Spain and Greece have long offered comparatively generous access to courts for environmental NGOs, at the same time as the Commission's annual monitoring reports have consistently singled them out as among the worst performers in implementing EU environmental law. Current efforts to increase access to courts are therefore unlikely to solve implementation problems there.

In part, obstacles to private enforcement in these countries lie in deeper problems with the effectiveness of the judicial system, its lack of resources and the training of judges (Moreno Molina 2012, p. 18; Kallia-Antoniou 2013, p. 52). However, an effective judicial system is a necessary condition for functioning private enforcement. The EU has little competence to address these issues beyond providing funds for networks of legal professionals. The recent inclusion of an 'EU Justice Scoreboard' in the European Semester, the EU's framework for monitoring socio-economic developments in member states, may at least map these issues (European Commission 2017b), but these examples show that the EU's ability to enhance national opportunity structures for private enforcement has very concrete limits.

What is more, the private enforcement of environmental law is essentially dependent on organised civil society showing an interest in the matter (Versluis 2007). Private enforcement is more likely to work in countries with more established and better funded environmental movements than in countries where such movements are young or largely absent (Börzel 2006; Slepcevic 2009). Legal mobilisation is also conditioned by environmental NGOs' alternative channels of influence within a country's political system or their ideological disposition towards law and courts as appropriate venues (Conant *et al.* 2017). There are significant differences in the use of courts between NGOs active in different sectors of environmental protection; bird protection NGOs have proven particularly litigious (Vanhala 2017). A primary reliance on de-centralised enforcement may improve the available remedies, but comes at the cost of significant imbalances both between member states and environmental sub-fields.

Accordingly, while the expansion of private access to courts is real, it is also fundamentally uneven, in ways that the EU has only limited capacity to address. Private enforcement can therefore not act as a universal substitute for Commission enforcement, and the Commission's continued presence is necessary to counter imbalances in private enforcement. There should be no illusions about the efficacy of either mechanism. Both have a role to play. The expansion of private access to courts was not the only cause of Commission retrenchment, but appears to give the Commission an excuse to do less. Alternatives to Commission enforcement, such as the delegation of direct monitoring, investigation and sanctioning powers to a specialised EU agency, as discussed by Collins and Earnshaw (1992, pp. 238–242), and which increasingly takes place in other policy areas (Scholten 2017), have so far failed to materialise in the environmental sector. The environmental sector forms part of a larger trend towards privatisation of enforcement (Wilman 2015, pp. 14–18). The result for the moment is not a balanced approach, but less and less activity on the part of the Commission. This ongoing retrenchment should be a cause for concern.

Acknowledgements

This contribution has benefitted greatly from discussions at the ECPR Joint Sessions workshop 'Whither the environment in Europe' in Pisa, 24–28 April 2016, and the workshop on 'The future of environmental policy in the European Union', hosted by the Centre for European Research (CERGU) at the University of Gothenburg, 19–20 January 2017. The latter workshop was funded by Riksbankens Jubileumsfond (Dnr F16-1249:1) and Vetenskåpsrådet (Dnr 2016-06930). I thank all participants for their input, in particular Charlotta Söderberg, Andrea Lenschow, Viviane Gravey, Lisa Vanhala and Chris Hilson, as well as the three anonymous referees.

Notes

1. The CJEU's case database at curia.europa.eu provides these numbers.
2. Calculation uses the Commission's database on infringement decisions at http://ec.europa.eu/atwork/applying-eu-law/infringements-proceedings/infringement_decisions.
3. The Commission's database on infringement decisions allows differentiation between non-notification cases and cases of substantive non-compliance.
4. Djurgården-Lilla Värtans Miljöskyddsförening stated that it had about 300 members.

Disclosure statement

No potential conflict of interest was reported by the author.

ORCID

Andreas Hofmann http://orcid.org/0000-0002-2014-6547

References

Bondarouk, E. and Mastenbroek, E., 2017. Reconsidering EU compliance: implementation performance in the field of environmental policy. *Environmental Policy and Governance*. doi:10.1002/eet.1761

Borrass, L., Sotirov, M., and Winkel, G., 2015. Policy change and Europeanization: implementing the European Union's Habitats Directive in Germany and the United Kingdom. *Environmental Politics*, 24 (5), 788–809. doi:10.1080/09644016.2015.1027056

Börzel, T.A., 2006. Participation through law enforcement: the case of the European Union. *Comparative Political Studies*, 39 (1), 128–152. doi:10.1177/0010414005283220

Börzel, T.A., et al. 2010. Obstinate and inefficient: why member states do not comply with European law. *Comparative Political Studies*, 43 (11), 1363–1390. doi:10.1177/0010414010376910

Börzel, T.A. and Buzogany, A., 2019. Compliance with EU environmental law. The iceberg is melting. *Environmental Politics*, 28 (2).

Börzel, T.A., Hofmann, T., and Panke, D., 2012. Caving in or sitting it out? Longitudinal patterns of non-compliance in the European Union. *Journal of European Public Policy*, 19 (4), 454–471. doi:10.1080/13501763.2011.607338

Brakeland, J.-F., 2014. Access to justice in environmental matters – developments at EU level. *Gyoseiho-kenkyu*, 2014(5). Available from: http://greenaccess.law.osaka-u.ac.jp/wp-content/uploads/2014/05/arten-brakelandup.pdf [Accessed 18 February 2018].

Bundesverwaltungsgericht, 2013. Judgment in case BVerwG 7 C 21.12, ECLI:DE: BVerwG:2013:050913U7C21.12.0.

Bunge, T., 2016. Weiter Zugang zu Gerichten nach der UVP- und der Industrieemissions-Richtlinie: vorgaben für das deutsche Verwaltungsprozessrecht. *Natur und Recht*, 38 (1), 11–21. doi:10.1007/s10357-015-2943-1

Čavoški, A., 2015. A post-austerity European Commission: no role for environmental policy? *Environmental Politics*, 24 (3), 501–505. doi:10.1080/09644016.2015.1008216

CJEU, 2008. Judgment of the Court in case C-237/07, Janecek, ECLI:EU:C:2008:447.

CJEU, 2009. Judgment of the Court in case C-263/08, Djurgården-Lilla Värtans Miljöskyddsförening, ECLI:EU:C:2009:631.

CJEU, 2011a. Judgment of the Court in case C-115/09, Bund für Umwelt und Naturschutz Deutschland, ECLI:EU:C:2011:289.

CJEU, 2011b. Judgment of the Court in case C-240/09, Lesoochranárske zoskupenie, ECLI:EU:C:2011:125.

CJEU, 2013. Judgment of the Court in case C-260/11, Edwards, ECLI:EU: C:2013:221.

CJEU, 2014. Judgment of the Court in case C-530/11, Commission v United Kingdom, ECLI:EU:C:2014:67.

CJEU, 2015. Judgment of the Court in case C-137/14, Commission v Germany, ECLI:EU:C:2015:683.

Clean Air, 2017. Lawsuits and decisions. Accessible from: http://legal.cleanair-europe.org/legal/germany/lawsuits-and-decisions [Accessed 18 February 2018].

Collins, K. and Earnshaw, D., 1992. The implementation and enforcement of European Community environment legislation. *Environmental Politics*, 1 (4), 213–249. doi:10.1080/09644019208414052

Conant, L., et al., 2017. Mobilizing European law. *Journal of European Public Policy*. doi:10.1080/13501763.2017.1329846

Council, 1973. Declaration of the Council of the European Communities on the programme of action of the European Communities on the environment, OJ C112, 20.12.1973, p. 1.

Council, 1993. Towards Sustainability. The European Community Programme of policy and action in relation to the environment and sustainable development, OJ C138, 17.5.1993, p. 5.

Craig, P. and de Búrca, G., 2011. *EU law. Text, cases and materials*. 5 ed. Oxford: Oxford University Press.

Darpö, J., 2012. Sweden: report of the study on the implementation of articles 9.3 and 9.4 of the Aarhus Convention in the member states of the European Union. Available from: http://jandarpo.se/wp-content/uploads/2017/07/EC-SE-Darpo.doc [Accessed 19 February 2008].

Darpö, J., 2013. Effective justice? Synthesis report of the study on the implementation of Articles 9.3 and 9.4 of the Aarhus Convention in the member states of the

European Union. Available from: http://uu.diva-portal.org/smash/get/diva2:694403/fulltext01.pdf [Accessed 19 February 2018].
Darpö, J., 2014. Med lagstiftaren på åskådarplats. Om implementeringen av Århuskonventionen genomrättspraxis. InfoTorg Juridik-Rättsbanken 2014-03-05. Available from: http://jandarpo.se/wp-content/uploads/2017/07/2014-RB-Anok.pdf [Accessed 20 February 2018].
De Baere, G. and Nowak, J.T., 2016. The right to "not prohibitively expensive" judicial proceedings under the Aarhus Convention and the ECJ as an international (environmental) law court: edwards and Pallikaropoulos. *Common Market Law Review*, 53, 1727–1752.
Directive 2003/35/EC of the European Parliament and of the Council providing for public participation in respect of the drawing up of certain plans and programmes relating to the environment, OJ L156, 5 6 2003, p. 17.
Directive 2003/4/EC of the European Parliament and of the Council on public access to environmental information, OJ L41, 14 2 2003, p. 26.
Eliantonio, M., 2015. The proceduralisation of EU environmental legislation: international pressures, some victories and some way to go. *Review of European Administrative Law 2015-1*, 8 (1), 99–123. doi:10.7590/187479815X14313382198377
Eliantonio, M., 2016. Enforcing EU environmental policy effectively: international influences, current barriers and possible solutions. *In*: S. Drake and M. Smith, eds.. *New directions in the effective enforcement of EU law and policy*. Cheltenham: Edward Elgar, 175–201.
Eliantonio, M. and Muir, E., 2015. Concluding thoughts: legitimacy, rationale and extent of the incidental proceduralisation of EU law. *Review of European Administrative Law*, 8 (1), 177–204. doi:10.7590/187479815X14313382198494
Epstein, Y. and Darpö, J., 2013. The wild has no words: environmental NGOs empowered to speak for protected species as Swedish courts apply EU and international environmental law. *Journal for European Environmental & Planning Law*, 10 (3), 250–261. doi:10.1163/18760104-01003004
European Commission, 1993. Tenth annual report to the European Parliament on Commission monitoring of the application of Community law. COM(93) 329 final.
European Commission, 1995. Twelfth annual report on monitoring the application of Community law. COM(95) 500 final.
European Commission, 1996. Communication from the Commission. Implementing Community environmental law. COM(96) 500.
European Commission, 2003. Proposal for a Directive of the European Parliament and of the Council on access to justice in environmental matters. COM(2003) 624 final.
European Commission, 2007. A Europe of results – applying Community law. COM(2007) 502 final
European Commission, 2011. Second evaluation report on EU Pilot. COM(2011) 930 final.
European Commission, 2014. Fitness check mandate for nature legislation. Available from: http://ec.europa.eu/environment/nature/legislation/fitness_check/docs/Mandate%20for%20Nature%20Legislation.pdf [Accessed 20 February 2018].
European Commission 2015a. Better regulation for better results - An EU agenda. COM(2015) 215 final.

European Commission, 2015b. Monitoring the application of Union law. 2014 Annual Report. COM(2015) 329 final.
European Commission, 2016a. Communication from the Commission. EU law: better results through better application. C(2016)8600.
European Commission, 2016b. Monitoring the application of Union law. 2015 Annual Report. COM(2016) 463 final.
European Commission, 2017a. Commission notice on access to justice in environmental matters. C(2017) 2616 final.
European Commission, 2017b. Communication from the Commission. The 2017 EU Justice Scoreboard. COM(2017) 167 final.
European Commission, 2017c. Communication from the Commission. The EU Environmental Implementation Review: common challenges and how to combine efforts to deliver better results. COM(2017) 63 final.
European Environmental Bureau, 2015. *EEB Work Programme 2016*. Brussels: EEB.
Falkner, G., et al. 2004. Non-compliance with EU directives in the member states: opposition through the backdoor? *West European Politics*, 27 (3), 452–473. doi:10.1080/0140238042000228095
Förvaltningsrätten i Luleå, 2011. Judgment in case 446-11.
Goodman, M. and Connelly, J., 2018. The public interest environmental law group: from USA to Europe. *Environmental Politics*, 27, 6. doi:10.1080/09644016.2018.1438789
Hedemann-Robinson, M., 2010. Enforcement of EU environmental law and the role of interim relief measures. *European Energy and Environmental Law Review*, 2010 (October), 204–229.
Hilson, C., 2002. New Social Movements: the role of legal opportunity. *Journal of European Public Policy*, 9 (2), 238–255. doi:10.1080/13501760110120246
Hilson, C., 2017. The visibility of environmental rights in the EU legal order: eurolegalism in action? *Journal of European Public Policy*. doi:10.1080/1350 1763.2017.1329335
Högsta Förvaltningsdomstolen, 2012. Judgment in case 2687-12.
Högsta Förvaltningsdomstolen, 2014. Judgment in case 5962-12.
IMPEL, 2000. Complaint procedures and access to justice for citizens and NGOs in the field of the environment within the European Union. Final report. Available from: https://www.elaw.org/system/files/E.U.-IMPEL.Network (Complaint.procedures.and.Access.to.Justice.for.citizens.and.NGO).pdf [Accesses 20 February 2018].
IMPEL, 2015. Challenges in the practical implementation of EU environmental law and how IMPEL could help overcome them. Final report. Available from: http://impel.eu/wp-content/uploads/2015/07/Implementation-Challenge-Report-23-March-2015.pdf [Accessed 20 February 2018]
Kallia-Antoniou, A., 2013. Study on factual aspects of access to justice in relation to EU environmental law: Greece. Available from: http://ec.europa.eu/environment/aarhus/pdf/studies2.zip [Accessed 20 February 2018]
Kassim, H., et al. 2017. Managing the house: the Presidency, agenda control and policy activism in the European Commission. *Journal of European Public Policy*, 24 (5), 653–674. doi:10.1080/13501763.2016.1154590
Kitschelt, H.P., 1986. Political opportunity structures and political protest: antinuclear movements in four democracies. *British Journal of Political Science*, 16 (1), 57–85. doi:10.1017/S000712340000380X

Krämer, L., 2014. EU enforcement of environmental laws: from great principles to daily practice – improving citizen involvement. *Environmental Policy and Law*, 44 (1–2), 247–256.

Lenaerts, K. and Gutiérrez-Fons, J.A., 2011. The general system of EU environmental law enforcement. *Yearbook of European Law*, 30 (1), 3–41. doi:10.1093/yel/yer001

Milieu, 2007. Summary report on the inventory of EU Member States' measures on access to justice in environmental matters. Available from: http://ec.europa.eu/environment/aarhus/pdf/studies.zip [Accessed 20 February 2018]

Moreno Molina, A.-M., 2012. Study on aspects of access to justice in relation to EU environmental law – the situation in Spain. Available from: http://ec.europa.eu/environment/aarhus/pdf/studies2.zip [Accessed 20 February 2018].

Oliver, P., 2013. Access to information and justice in EU environmental law. *Fordham International Law Journal*, 36 (4), 1423–1470.

Poncelet, C., 2012. Access to justice in environmental matters. Does the European Union comply with its obligations? *Journal of Environmental Law*, 24 (2), 287–309. doi:10.1093/jel/eqs004

Pressman, J.L. and Wildavsky, A.B., 1974. *Implementation. How great expectations in Washington are dashed in Oakland*. Berkeley: University of California Press.

Prieur, M., 1998. Complaints and appeals in the area of environment in the member states of the European Union. General report. Study for the Commission of the European Community, Brussel.

Ryall, A., 2013. Study on the implementation of article 9 (3)and 9 (4)of the Aarhus Convention in 17 of the member states of the European Union. Ireland. Available from: http://ec.europa.eu/environment/aarhus/pdf/studies2.zip [Accessed 20 February 2018].

Scholten, M., 2017. Mind the trend! Enforcement of EU law has been moving to 'Brussels'. *Journal of European Public Policy*. doi:10.1080/13501763.2017.1314538

Slepcevic, R., 2009. The judicial enforcement of EU law through national courts: possibilities and limits. *Journal of European Public Policy*, 16 (3), 378–394. doi:10.1080/13501760802662847

Smith, M., 2016. The visible, the invisible and the impenetrable: innovations or rebranding regulatory goals and constituitional values. *In*: S. Drake and M. Smith, eds. *New directions in the effective enforcement of EU law and policy*. Cheltenham: Edward Elgar, 45–76.

Swedish Code of Statutes, 2010. Lag om ändring i miljöbalken, SFS 2010:882. Available from: http://www.notisum.se/kbvlag/20100882.pdf [Accessed 20 February 2018]

Tallberg, J., 2002. Paths to compliance: enforcement, management, and the European Union. *International Organization*, 56 (3), 609–643. doi:10.1162/002081802760199908

Thomann, E. and Sager, F., 2017. Moving beyond legal compliance: innovative approaches to EU multilevel implementation. *Journal of European Public Policy*. doi:10.1080/13501763.2017.1314541

Treib, O., 2014. Implementing and complying with EU governance outputs. *Living Reviews in European Governance*, 9 (1), 1–47. doi:10.12942/lreg-2014-1

Vanhala, L., 2017. Is legal mobilization for the birds? Legal opportunity structures and environmental nongovernmental organisations in the United Kingdom, France, Finland and Italy. *Comparative Political Studies*, 50 (3), 380–412.

Versluis, E., 2007. Even rules, uneven practices: opening the 'black box' of EU law in action. *West European Politics*, 30 (1), 50–67. doi:10.1080/01402380601019647

Wilman, F., 2015. *Private enforcement of EU law before national courts. The EU legislative framework*. Cheltenham: Edward Elgar.

WWF, 2015a. Nature alert: environmental groups rally as European Commission threatens vital nature laws. Press release. Available from: http://www.wwf.eu/?246530/Nature-alert-environmental-groups-rally-as-European-Commission-threatens-vital-nature-laws [Accessed 20 February 2018].

WWF, 2015b. *WWF EPO Annual Review 2014*. Brussels: WWF.

Zito, A., Burns, C., and Lenschow, A., 2019. Is the trajectory of European Union environmental policy less certain? *Environmental Politics*, 28 (2).

ə OPEN ACCESS

Environmental policy evaluation in the EU: between learning, accountability, and political opportunities?

Jonas J. Schoenefeld and Andrew J. Jordan

ABSTRACT
Policy evaluation has grown significantly in the EU environmental sector since the 1990s. In identifying and exploring the putative drivers behind its rise – a desire to learn, a quest for greater accountability, and a wish to manipulate political opportunity structures – new ground is broken by examining how and why the existing literatures on these drivers have largely studied them in isolation. The complementarities and potential tensions between the three drivers are then addressed in order to advance existing research, drawing on emerging empirical examples in climate policy, a very dynamic area of evaluation activity in the EU. The conclusions suggest that future studies should explore the interactions between the three drivers to open up new and exciting research opportunities in order to comprehend contemporary environmental policy and politics in the EU.

Introduction

In the 25 years since *Environmental Politics* published its seminal special issue on European Union (EU) environmental policy (Judge 1992), policy evaluation (hereafter 'evaluation') has flourished. Our contribution seeks to identify the core drivers that lie behind the EU's increasing proclivity to evaluate its environmental policies. Doing so matters because the resources committed to evaluation are substantial. In 2007, the European Commission employed 140 full-time staff in this area and spent 45 million Euros (Højlund 2015, p. 36). These investments are generating significant outputs in the form of new, policy-relevant knowledge. Mastenbroek *et al.* (2016) found that the European Commission initiated 216 *ex-post* legislative evaluations between 2000 and 2012, with significant growth in more recent years: 'nearly 200 evaluations were published [by the Commission] … between January 2015 and mid-October 2016 […]' Schrefler (2016, p. 6). These numbers only

This is an Open Access article distributed under the terms of the Creative Commons Attribution License (http://creativecommons.org/licenses/by/4.0/), which permits unrestricted use, distribution, and reproduction in any medium, provided the original work is properly cited.

partially represent total evaluation output, given that other institutions, such as the European Court of Auditors (Stephenson 2015) and the European Parliament, as well as non-governmental organisations and industry associations, also evaluate (see Schoenefeld and Jordan 2017). The European Environment Agency (EEA) recently wrote that '[t]he evaluation of environment and climate policies is, today, a well-established discipline' (2016, p. 4). Textbooks on EU environmental policy now incorporate chapters on evaluation (e.g. Mickwitz 2013); a meta-analysis conducted at a time when climate policy outputs were growing rapidly, found over 250 evaluations in the sub-area of climate policy (Haug et al. 2010, Huitema et al. 2011).

While evaluation has become an established feature of EU environmental policymaking, it is much less clear *why* this has occurred. What are the key motivations of those engaging in evaluation? And to what extent can evaluation fulfil their aspirations? We follow Vedung (1997, p. 3) in defining evaluation as a 'careful retrospective assessment of the merit, worth and value of administration, output and outcome of government interventions, which is intended to play a role in future, practical action situations'. The key word is 'retrospective'; we focus here on *ex-post* evaluations (see also Mickwitz 2006, Crabbé and Leroy 2008), rather than on their *ex-nunc* (ongoing – see Crabbé and Leroy 2008) or *ex-ante* (prospective) elements (see Adelle et al. 2012, Turnpenny et al. 2016).

Until now, evaluation scholars have mainly concentrated on developing evaluation methods (e.g. Vedung 1997, Pawson and Tilley 2014), including on environmental policy (Mickwitz 2003, Crabbé and Leroy 2008). Work exploring the underlying drivers of evaluation has emerged only quite recently. Even though the 1992 special issue considered implementation (Collins and Earnshaw 1992), it did not address evaluation. Very few scholars have worked specifically on environmental evaluation in the EU (but see Mickwitz 2013).

The general neglect of evaluation matters because there are multiple reasons why actors may advocate, commission, fund, undertake, enact, and/or respond to evaluation. Scholars such as Radaelli (2010) and Adelle *et al.* (2012) have explored the politics of *ex-ante* impact assessment, but this has been much less the case for *ex-post* evaluation. Another key shortcoming in the *ex-post* evaluation literatures is that few scholars have considered multiple evaluation drivers together. Most existing accounts analyse the drivers in isolation (e.g. Bovens et al. 2006). For example, even though the prominent evaluation scholar Elinor Chelimsky asserts that '[m]y point is that claiming a unique purpose for evaluation flies in the face of past and current practice' (2006, p. 36), she neglects political aspects in her own review of the field. Our core aim is to incorporate all three drivers of evaluation, namely: a quest for learning; an enabler of accountability; and a way to manipulate political opportunity structures.

We proceed as follows. The next section reviews the emergence of evaluation (and especially environmental evaluation) in the EU. It focuses on evaluation's role in the EU Environmental Action Programmes (EAP), which the EU publishes regularly in order to guide and frame its environmental policy work. Their strategic nature makes them a suitable indicator of deeper shifts in EU environmental policy-making (see Mickwitz 2013). The third section returns to the three evaluation drivers outlined above and explores them theoretically, drawing on new empirical insights which are beginning to appear in the literature. The fourth section conceptualises the interaction between the drivers, drawing on emerging empirical evidence. The fifth draws together the main findings, concludes, and identifies new research needs.

Emergence of environmental policy evaluation in the EU

Most histories of evaluation identify its origins in the USA where actors assessed social policy in the 1960s (Toulemonde 2000, Stame 2003). About two decades later, the rise of New Public Management, which aims at more efficient and effective policy-making, proved influential in popularising evaluation in Europe (Pattyn 2014, Pattyn et al. 2018). Other factors include EU enlargement, and a perceived need to evaluate the effectiveness of structural and cohesion funding that was increasingly being dispersed east and southwards (e.g. Batterbury 2006) as well as encouragement from the OECD and the World Bank (Toulemonde 2000, Uitto 2016). More recently, scholars have also noted 'better regulation' initiatives and concerns over policy effectiveness in a world of dwindling public budgets as potential drivers (see EEA 2016).

Environment and climate change evaluation only emerged in the mid-1990s in the EU – largely following similar earlier trends in the USA (Knaap and Kim 1998, p. 23, see also Feldman and Wilt 1996). One reason for this lag may be that, as Toulemonde writes, 'professional [evaluation] networks have remained highly compartmentalised and hardly inclined to bridge the gap with other sectors' (2000, p. 351). Professional evaluators have typically focused on the EU fields where evaluation first developed, notably structural funds and research policy. However, policymakers were equally slow to demand evaluations of environmental policy. While evaluation first centred on spending policies, environmental policy was, and to a large extent remains, a regulatory affair in order to avoid distortions in the common market (Knill and Liefferink 2013) and has thus often been subjected to less evaluation. Regulation tends to be less political because the benefits it generates tend to be diffuse and slow to appear (see Majone 1994). It also took time for the environmental *acquis* to expand enough to generate effects that demanded evaluating (as happened in the USA, where environmental evaluation only really emerged a decade or so after significant legislation had been adopted – see Knaap and Kim 1998, p. 23).

However, EU environmental policy could not escape these broader trends forever (e.g. Toulemonde 2000, Mickwitz 2006, 2013, Stame 2008, EEA 2016). Evaluation did not suddenly appear in the environmental sector; rather, it gradually built over time. Some of its origins lie in earlier practices such as regulatory impact assessment. To trace this development, it is worth exploring evaluation's rising prominence in the EU's EAPs over time. These programmes identify strategic priorities for EU environmental policy, including in evaluation (see Mickwitz 2006). Table 1 summarises the appearance of 'assessment' and 'evaluation' in the seven EAPs to date.

Table 1 reveals that references to policy assessment date back to the first EAP, but have strengthened and become more common over time. The excerpts reveal how successive EAPs have defined assessments more concretely with more specific language on methodology. The focus has evolved from

Table 1. Evaluation in the EU's environmental action programmes.

EAP Number & Year	References to evaluation
1st (1973)	The protection of the environment [...] inevitably involves various kinds of expenditure [...] It is essential that the authorities make accurate assessments of the size of this expenditure in order to have a clear idea of what the **economic, financial and social repercussions of proposed decisions are likely to be[...]** (p. 31; emphasis added)
2nd (1977)	The Commission will **try to find a method** of costing anti-pollution measures... (p. 37; emphasis added)
3rd (1983)	**Environmental impact assessment** is the prime instrument for ensuring that environmental data is taken into account in the decision-making process. (p. 6; emphasis added)
4th (1987)	[...] Community environment actions shall take account of the potential benefits and costs of action or of lack of action. The Commission will **endeavour to develop methods of assessment** which will facilitate this task and which will, so far as possible, ensure the preparation of an adequate cost benefit analysis as a basis for environmental proposals. (p. 14; emphasis added)
5th (1993)	[...] the design and the choice of environmental priorities must be elaborated, **based on the fullest possible assessment of all relevant costs and benefits**. (p. 97; emphasis added)
6th (2002)	[...] improvement of the process of policy making through: – **ex-ante evaluation** of the possible impacts, in particular the environmental impacts, of new policies including the alternative of no action and of the proposals for legislation and publication of the results; – **ex-post evaluation** of the effectiveness of existing measures in meeting their environmental objectives [...] (p. 13–14; emphasis added)
7th (2013)	In order to improve environmental integration and policy coherence, the 7th EAP shall ensure that by 2020: [...] This requires, in particular: (i) integrating environmental and climate-related conditionalities [...] in policy initiatives, including **reviews and reforms of existing policy**, as well as new initiatives, at Union and Member State level; (ii) carrying out **ex-ante assessments** of the environmental, social and economic impacts of policy initiatives [...] to ensure their coherence and effectiveness; [...] (iv) using **ex-post evaluation information** relating to experience with implementation of the environment acquis in order to improve its consistency and coherence [...] (p. 195–196; emphasis added)

environment-related expenditure to incorporating economic effects, including costs and benefits, and, from 1987, the costs of inaction. However, explicit references to *ex-post* evaluation only emerged in the 6th and 7th EAPs (see also Mickwitz 2013), which was about ten years after evaluation became a standard part of the policy repertoire in the USA. A 1996 Communication by the Commission sought to systematise evaluation practices in the EU (European Commission 1996). Recent research shows that compared to other policy areas, DG Environment initiated a moderate level (around 40) of legislative *ex-post* evaluations between 2000 and 2014 (van Voorst and Mastenbroek 2017). Evaluation practices have thus grown rapidly in the EU since the late 1990s, often in the absence of a clear blueprint from the Commission or the Member States. But what are the deeper drivers of this trend? The following section identifies and explores three core drivers in detail.

Evaluation drivers: existing debates

Academic literatures on EU evaluation have, over time, consistently and repeatedly stressed three underlying drivers of evaluation. Two drivers that often feature in evaluation debates are accountability and learning – the latter often starting from the idea of evaluation as the last 'stage' in a stylised 'policy cycle' (Hanberger 2012, Vo and Christie 2015). However, actors may use evaluation in order to manipulate political opportunity structures, from using evaluation to delay processes through to legitimising pre-existing policy actions (Hanberger 2012). This section assesses these debates with a view to identifying what we know about the three drivers (see also Vedung 1997, p. 13).

Accountability

Many existing literatures focus on evaluation as an accountability mechanism. Bovens (2010) explains that meanings of accountability incorporate normative visions of transparency and virtue, and potentially organisational mechanisms through which agents answer to their principals – an important issue for climate change policy, which often involves numerous actors at various governance levels (Feldman and Wilt 1996, Jordan *et al.* 2015, 2018, Schoenefeld and Jordan 2017). In seeking to link evaluation and accountability, the relevant literatures mainly focus on the latter function, envisaging evaluation as an enabler of accountability (Stame 2003, Hanberger 2012) through processes of policy surveillance (see Aldy 2014). As Alkin and Christie assume: '[t]he need and desire for accountability presents a need for evaluation' (2004, p. 12). To fulfil this role, many scholars emphasise the need for 'independent' evaluations that are removed from the turmoil of everyday politics (Weiss 1993, Feldman and Wilt 1996). For example, Chelimsky envisions evaluation as being largely external to government, emphasising that '[a]fter all, evaluation exists to report on government, not to be a part of it' (2009, p. 65). Relatedly, Hildén (2011) stresses that powerful governmental actors may constrain government-sponsored or -produced evaluations. Taken together, evaluation may support key accountability mechanisms within states (Hanberger 2012).

However, there is a growing recognition that, particularly in an EU context, hierarchical, state-like structures have, in part, given way to more networked (e.g. Rhodes 1996) and, especially in the case of climate change, increasingly polycentric governance arrangements (see Dorsch and Flachsland 2017, Jordan et al. 2018). High levels of complexity and multiple actors in environmental governance make it especially difficult to ascertain who should be held accountable for which policy outcomes (van der Meer and Edelenbos 2006); a very politicised activity since holding organisations accountable for their actions is highly visible compared to, for example, the potentially subtler politics within *ex-ante* assessment of exploring *potential* impacts. New forms of accountability have thus emerged, such as horizontal accountability to a range of actors, including civil society (Bovens 2007, Hertting and Vedung 2012). This has profound implications for evaluation: if there is not one but many principals, evaluation may require broader approaches and multiple criteria (Hanberger 2012), as well as the involvement of numerous stakeholders (Hertting and Vedung 2012). Recent debates have thus focused on multiple criteria and triangulation in EU environmental policy evaluation (Mickwitz 2013). Scholars have often envisioned evaluation as an enabler of accountability in state-like and increasingly networked governance through knowledge provision. Some even argue that evaluation enables democratic processes by stimulating debate (Toulemonde 2000, Stame 2006).

Thus far we have discussed a normative case for evaluation as an accountability mechanism, but what do we know about the extent to which such accountability functions actually materialise in the EU? Recent evidence casts some doubt on these optimistic, normative visions for evaluation as an accountability mechanism. In a study of 220 legislative evaluations, Zwaan et al. (2016) found that only 16% were discussed in the European Parliament; even then, the main motivation appears to have been agenda-setting rather than holding the European Commission to account. While their study only considers evaluations carried out for the European Commission, it points to the need to further empirically investigate the assumed accountability functions of evaluation.

Learning

The second commonly discussed evaluation driver is policy improvement or policy learning. However, does a desire to learn actually stimulate evaluation? Much like accountability, policy learning is a contested concept, involving many different forms (Zito and Schout 2009). In turn, these often arise from different perspectives on the nature of EU governance and its functions (see Radaelli and Dunlop 2013). Evaluation is often assumed to deliver critical inputs to stimulate learning (the so-called objectivist view) or to facilitate a process through which participants learn (the more argumentative view) (Borrás and Højlund 2015; see also Hildén 2011). Thus, as Haug argues, '[e]x-post evaluation of programmes or policies [...] is a widely applied group of approaches aimed at stimulating learning in environmental governance' (2015, p. 5). Hanberger (2012) develops the objectivist view to propose that a more hierarchical state-like organisation would benefit from information on policy effectiveness, whereas more network-like settings require evaluation that focuses on how collaboration works. States would thus learn from their elites using evaluation, while networks would learn from collective processes that evaluation enables (Hanberger 2012). This demonstrates that many evaluation and governance literatures still envision learning as a more or less direct 'feedback loop', meaning that actors learn through the knowledge they receive from evaluation. Relatedly, the WWF argued in its Climate Policy Tracker for the EU that '[t]he evaluation of past performance is important to verify the effectiveness and efficiency of measures, to learn about their driving forces and adjust policies accordingly' (2010, p. 16). This goes hand-in-glove with a rational 'evidence-based' policy-making view (for a fuller discussion, see Sanderson 2002).

Nevertheless, do such normative beliefs about evaluation's role in learning actually materialise in practice, or are references to learning largely rhetorical devices in order to justify evaluation done for other, more political, reasons? Long ago, evaluation scholars realised that the direct, linear use of evaluation is extremely rare (Weiss 1999). They stress how learning works through more nuanced and indirect processes (Zito and Schout 2009), such as Weiss's (1999) 'enlightenment', or situations where evaluation and monitoring exercises may perform more of a 'radar' function (Radaelli and Dunlop 2013). This is at least in part because evaluation is by no means the only source of knowledge and pressure on policy-makers (Weiss 1999). Challengingly, learning as improvement vis-à-vis evaluation ultimately requires consensus on policy values meaning that things are 'improving' on dimensions that particular actors deem relevant, important, and thus worthy of action.

This state of affairs generates at least two pertinent research questions: first, what do we know about the extent to which environmental evaluations facilitate learning? Focusing on 'government learning', Borrás and Højlund (2015) investigated the learning arising from three evaluations commissioned by the European Commission (two focusing on environmental policy). Based on intensive interview research, they found that learning did take place, with programme or unit officers and external evaluators among the prime learners. It ranged from gaining a fuller overview of the policy area to learning about new evaluation methodologies, although interviewees stressed the incremental nature of both processes (Borrás and Højlund 2015). Focusing on climate policy in Finland, Hildén (2011) highlighted multiple forms of learning, but also detected political and rhetorical learning.

Together, these findings indicate that evaluation may indeed contribute to learning at EU level, but more research is required to assess what kind and under what conditions learning occurs as a function of evaluation. Second, it is pertinent to ask why the simplistic, linear view of learning appears to persist among evaluation scholars and especially practitioners, and what may be the alternative (e.g. more political) drivers of evaluation? The next section addresses these questions.

Political opportunity structures

Actors also use evaluation in order to manipulate political opportunity structures (McAdam 1996) as part of much broader political struggles (Bovens *et al.* 2006, Weiss 1993, Vedung 1997). Evaluation may expand or reduce the 'scope of political conflict' (Schattschneider 1975) by bringing certain actors into policy discussions, for example, through direct participation in an evaluation as a 'stakeholder' or by using evaluation results in public debates (see the introduction to this special issue, Zito *et al.* 2019). The more polycentric climate governance 'opportunity structure' emerging from the Paris Agreement (UNFCCC 2015) and its application in the EU (see Tosun and Schoenefeld 2017, Ringel and Knodt 2018) is likely to expand such access points. However, evaluations can also exclude actors or end discussion by delegating debates to evaluators or by effectively delaying political processes (Pollitt 1998). For some actors, engagement in evaluation has little to do with enabling accountability or fostering learning through evaluation; rather, it is a way to manipulate opportunity structures in order to advance their political goals. We thus should not understand evaluation as a 'clinical, experimental science', but something that is part and parcel of wider political processes (see Weiss 1993).

At a fairly basic level, evaluation may allow certain actors to participate in governance processes, and potentially shut out others (and thus affect the participatory structure). Manipulating political opportunity structures may furthermore involve shifting power relations among actors (see Schoenefeld and Jordan 2017) by legitimising certain actions, actors, or ideas, and delegitimising others (Hanberger 2012). Actors may commission evaluations simply to appear legitimate (i.e. by claiming that their decisions are evidence-based), with little interest in the results of the exercise. Alternatively, they may use evaluations for political or symbolic, rather than more substantial, forms of learning. Actors may also utilise evaluation in order to avoid political conflict and/or escape blame (see Howlett 2014). The creation of evaluation units within the European Commission, and since 2014 a dedicated Commissioner for better regulation, have certainly sent a strong political signal by strengthening the institutional basis of evaluation. There are, however, other elements of evaluation that relate to the governance questions noted above: actors may decide to evaluate (or not) based on their pre-conceptions of policy success, or they may seek to influence the evaluation process so that the results suit pre-defined policy objectives (as a form of policy-based evidence). Policy-makers may even seek to legitimise certain policies by subjecting them to repeated evaluations.

A common response to these issues has often been to devise mechanisms that protect evaluators and their organisations from political pressure, for example, by creating independent evaluation units (Chelimsky 2009). There have thus been multiple attempts to organise the politics out of evaluation. These attempts to depoliticise completely evaluation have often been futile however, as evaluation always involves making value judgements (see Vedung 1997). Working towards a fuller understanding of manipulating political opportunity structures through evaluation requires an understanding of the various actors involved in pursuing, financing, commissioning, and/or conducting evaluations, and their core motivations. To date, our knowledge of actor motivation is at best patchy and at worst non-existent in the area of environment and climate policy in the EU (but see Schoenefeld and Jordan 2017). It is then also useful to understand the nature of the evaluation processes, the outputs and ideas they generate, and the usage of the outcomes – for example, in agenda-setting and policy formulation. In recent years, scholars have endeavoured to address these important empirical questions generally, and also with regard to environment and climate evaluation. This work details the growing evaluation activities in the European Commission as one key actor in pursuit of evaluation (e.g. Højlund 2015, Mastenbroek *et al.* 2016), but also increasingly the European Court of Auditors (e.g. Stephenson 2015), the EEA (EEA 2016, Schoenefeld *et al.* 2018), or across the EU as a whole (e.g. Stern 2009, Jacob *et al.* 2015).

Extant work has often focused on mapping the evaluation outputs of particular EU institutions. For example, Mastenbroek *et al.* (2016) found 216 studies focusing on *ex-post* legislative evaluation conducted by the European Commission and van Voorst and Mastenbroek (2017) have elaborated on it to test causal models. Warren (2014) conducted a meta-evaluation of experiences with demand-side energy policy. One of the key challenges of these literatures is that the sampling criteria for collecting evaluations vary widely, so that it is difficult if not impossible to compare their results, let alone explore the reasons for evaluation growth. For example, Huitema *et al.* (2011) included academic articles as 'evaluations' in their study, while Mastenbroek et al. (2016) only focused on evaluations by the European Commission. By contrast, Warren (2014) drew on academic databases in his analysis, but neglected evaluations published in other venues, such as those identified by Mastenbroek et al. (2016). In sum, working towards clearer concepts that can be operationalised is a key first step in advancing this field towards more causal explanations of its politics (and the other drivers).

An emerging and important line of research considers the relationship between those who commission evaluations and those who conduct them. In a survey, Hayward et al. (2014) showed how members of the British government frequently aimed to influence the evaluators that they had commissioned to conduct evaluations, whether through influencing their methodologies or during the final write-up; Pleger and Sager (2016) have discovered similar dynamics in Germany and Switzerland. Even though earlier studies demonstrate that EU-level actors frequently commission environment and especially climate policy evaluations (Huitema et al. 2011), these dynamics have not yet been sufficiently explored.

Next steps

A key shortcoming of the existing literatures reviewed above is that they have hardly considered the drivers of evaluation side-by-side, either theoretically or empirically, especially in the case of EU environmental policy. As a first step, this section begins to work across them theoretically, a key endeavour in order to enable new theory-driven, empirical explorations. Second, it looks at recent empirical work that has begun to lay bare potential overlaps, as well as tensions, between the drivers.

Working across the drivers theoretically

As a first step towards building a more comprehensive understanding, Figure 1 maps the theoretical concepts and identifies their main overlaps and/or tensions.

Figure 1 depicts each driver in a circle in order to identify tensions and/or overlap between them, drawing on existing evaluation literatures. The figure affords significant space to each of the three drivers in order to propose that it appears conceptually helpful to understanding them individually (i.e. we found no indication of a perfect overlap between any two drivers in the evaluation literatures). Equally, we identified areas with significant potential conceptual overlap between the drivers. For example, instances where 'learning' and 'manipulating political opportunity structures' occur simultaneously may lead to 'political learning'. Similarly, when accountability and manipulating political opportunity structures overlap, a theoretical result may become some form of

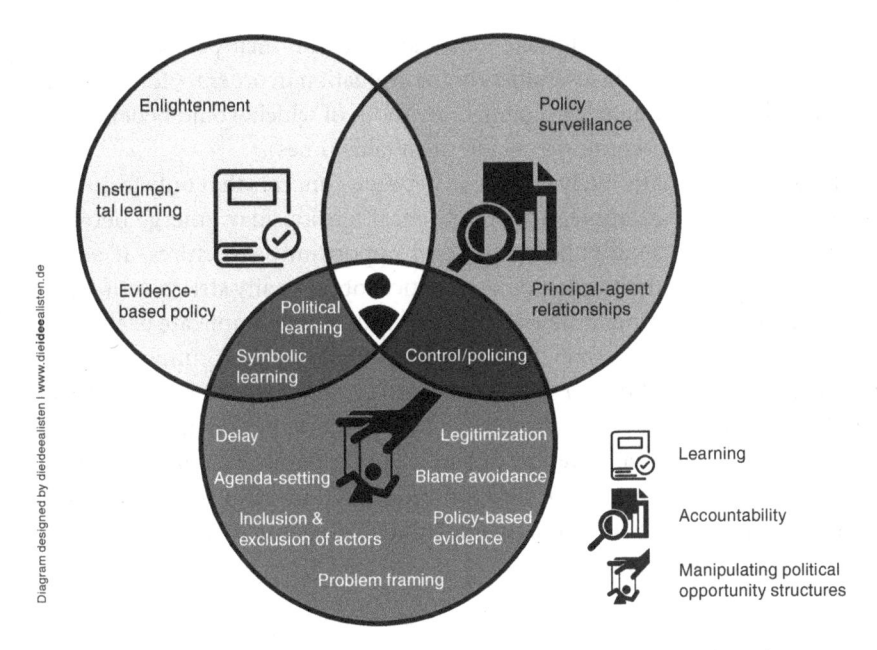

Figure 1. Learning, accountability, and political opportunities through evaluation.

policing or control. Last, a potential conceptual overlap between learning and accountability is less clear and points to tension between the two, but Regeer *et al.* (2016) argue for extending the concept of accountability in order to make learning a sub-set of it – but they acknowledge that evaluation focuses more on accountability than on learning. Stame (2003) writes that potential overlaps between accountability and learning depend on the relationship between principals and agents at the outset; if organisational goals are similar, the evaluation can serve a productive role, both in enabling accountability and potentially learning (complementarity – see also Sabel 1994, OECD 2001); however, if organisational goals differ, accountability functions may come at the cost of reduced or even no learning effects (antagonism). One concept that has been advanced to incorporate both (while arguable ignoring the potential tensions) is policy surveillance, which Aldy (2018, p. 211) argues includes

> reporting and monitoring of relevant climate policy performance data, as well as the analysis and evaluation of those data. Doing so can facilitate learning about the efficacy of mitigation efforts...

However, particularly where few concepts point to an overlap between the drivers in Figure 1, tensions may emerge that may prevent the simultaneous manifestation of the normative ideals articulated through the three evaluation drivers as a consequence of evaluation. This is because theoretically one could expect significant antagonism or tension between the drivers. For example, if an actor conducts an evaluation in order to delay political processes, they would want the evaluation to take long enough (from their perspective), and put lower emphasis on the usefulness of the evaluation in order to, for example, stimulate learning or enable accountability (both of which would benefit from evaluation insights becoming available at suitable times).

Tension is especially likely to emerge between concepts that only feature in a single circle. For example, a key theoretical tension may emerge between accountability and manipulating political opportunity structures. If actors use evaluation in order to manipulate political opportunity structures (e.g. to delay processes or bring in new actors) and thus deeply implicate evaluation with the related governance processes, it is all but impossible to conceptualise evaluation, *at the same time*, as an 'external' accountability mechanism. Ironically, evaluation may thus become subject to accountability pressures, such as when different organisations commission competing evaluations in order to support or delegitimise certain ideas. In sum, the extent to which these processes are antagonistic or complementary still requires more conceptual exploration.

Working across the drivers empirically

Exposing the theoretical and often deeply normative arguments outlined in the previous sections to empirical scrutiny is a key, but only partially realised, objective. Empirical analyses of interactions between the three drivers have often concentrated on just a few of the wide range of potential overlaps and/or tensions. The interstitials between learning and accountability attract most attention. Sabel (1994) has noted that monitoring could lead to learning if institutions forge mutual interest among various actors, who then come to understand an ongoing conversation about monitoring standards and outcomes as a learning opportunity. However, others argue that different evaluation approaches either suit learning or accountability: Højlund (2015), for example, explains that summative evaluations (at the end of a policy) are more accountability-oriented, whereas formative evaluations (occurring while a policy is being implemented) may be more geared towards learning. Tensions between accountability and learning may also emerge because evaluations can be used to name and shame governance actors, and they are effectively a control function, which can erode trust and a willingness to consider seriously potential improvements (Hermans 2009). Similarly, van der Meer and Edelenbos have highlighted that:

> Evaluations that primarily have an accountability function tend to be public and are often performed by external evaluators. Units whose policies, management or implementation activities are being evaluated will often be inclined to defend their actions and achievements. (2006, p. 209)

Højlund (2015) has documented how evaluation in the European Commission has oscillated between accountability and learning. More streamlining since the 1990s has usually meant a greater focus on accountability/control, and less on learning. While accountability and learning may not always be opposed to each other, more empirical evidence is needed to investigate their relationships. This is especially because high political hopes are currently investing in the accountability and learning functions of evaluation in the EU. For example, the Research Service of the European Parliament concludes that

> Evaluation is an *important element for the proper functioning of the policy cycle*. It serves many purposes, for instance *assessing how a particular policy intervention has performed in comparison with expectations* [...]. Evaluation is also a means of *fostering transparency and accountability* towards citizens and stakeholders. Last but not least, evaluation provides *evidence for policy-makers in deciding whether to continue, modify or terminate a policy intervention*. (Schrefler 2016, 5 – emphasis added)

The EEA has recently expressed similar expectations about environment and climate evaluation in the EU (EEA 2016), which is again a call to researchers to assess whether these hopes materialise. We contend that it is very much an open question – both theoretically and empirically – whether evaluation can and does fulfil these high expectations in the areas of environment and climate policy. The implicit assumption in the quote above is that evaluation can fulfil all these roles *at the same time*. There is little recognition of potential or real tensions between the accountability and learning functions of evaluation, or that evaluation could also serve as ammunition in political battles. The hopes in the quote thus contrast sharply with emerging theoretical debates and empirical evidence on these dynamics.

Work is also emerging on the overlap between manipulating political opportunity structures and accountability. Schoenefeld *et al.* (2018) show how EU Member States were reluctant to strengthen climate policy monitoring in 2013 lest it increased the power of EU level actors. As Stame (2003) writes, monitoring is a less 'intrusive' activity than evaluation – in the case of the latter, concerns over control may be even stronger. As Schoenefeld *et al.* (2018) discuss, this may have severe ramifications for learning, because patchy and insufficient knowledge as well as limited indicators have disabled broader climate governance debates. Finally, the overlap between manipulating political opportunity structures and learning with a view to improving EU environmental policy evaluation is ripe for deeper empirical exploration.

While exploring the three drivers in pairs is certainly helpful, ultimately it would be useful to explore empirically all three in conjunction; that is, exploring the central area in Figure 1. A good place to begin addressing this research challenge is the EEA, which is central to many environmental policy knowledge development and dissemination activities (Martens 2010). Whereas the European Commission and Member States have generally preferred the EEA to focus on data collection, the European Parliament prefers a stronger role in policy analysis in order to hold the Commission and the Council to account (Martens 2010). These tensions have certainly manifested in the EEA's approach to monitoring climate policies. A recent analysis of the outputs of the EU's Monitoring Mechanism on climate change – which the EEA manages – identified a range of political conflicts and tensions, ranging from Member State concern over reporting costs to fears of losing political control over knowledge generation and sharing and, potentially, even future target setting (Schoenefeld et al. 2018). By the same token, we know much less about the activities of non-governmental actors in evaluation (see Hildén et al. 2014).

There are other examples of new research that has worked across the three drivers. For instance, van Voorst and Mastenbroek (2017) empirically tested motivations for evaluation – including aspects of accountability, learning, and politics – finding that the Commission is most likely to evaluate in order to enforce legislation (hence more accountability than learning), and when evaluation capacities are high. They did not find that politicisation in the Council (i.e. the apex of decision-making and hence the most openly political level) affected the initiation and framing of particular evaluations.

Conclusions

Since the original special issue on EU environmental policy (Judge 1992), evaluation has become an important element of environmental policymaking in the EU, and hence should be accounted for in any attempt to take stock of EU environmental policy and governance. There has undoubtedly been a clear change at EU level, expressed by the steep growth in political support and demand for evaluation, resource investments, and evaluation outputs. This change has been gradual over time, with an initial international impetus in the 1960s, followed by a strengthening of evaluation-related language in the EU Environmental Action Programs. In the last ten years, the institutionalisation of evaluation has materialised with many more evaluations published, together with leadership from the relevant European institutions and new guidance on best practice. Drawing on various existing literatures, we argue that evaluation has emerged from three core underlying drivers: a quest to foster policy learning; a perceived need for greater accountability; and a desire to manipulate political opportunity structures. We unpacked each driver theoretically and explored its empirical relevance before making a first attempt to work across the three drivers theoretically. We then drew on the emerging literatures on evaluation in order to explore the extent to which some of the overlaps and/ or tensions between the drivers have been empirically explored.

High levels of complexity and uncertainty typically characterise environmental (and especially climate change) policy (Mickwitz 2013) and make the conceptualisation of accountability, learning, and politics especially challenging and related effects empirically hard to detect. This is especially because many environmental issues including climate change do not neatly coincide with existing political jurisdictions (Bruyninckx 2009). The nature of the field thus invites multiple evaluation approaches and diversity in evaluation (Mickwitz 2006). Unlike more mature areas of EU evaluation such as structural funding and international development in which it is abundantly clear who has an active interest in evaluation (the Member States, as donors), in the fields of environment and climate policy the overall picture is murkier (see Schoenefeld and Jordan 2017). While we have some knowledge about which actors have in the past advocated evaluations and for what reasons, there has been a marked reluctance to open up the associated political and institutional aspects to further scrutiny.

Future research is necessary in order to further disentangle the different drivers and especially, to assess their empirical relevance (e.g. does a rhetorical emphasis on accountability and/or learning functions of evaluation really materialise empirically, or are political factors more prominent?), and thereby test the relationships we proposed in Figure 1. Where can we identify further evidence of tension and/or overlap between the drivers and the sub-concepts included in the Figure? In the area of environmental policy it matters who is undertaking evaluation, where, when, why, how, for what reasons, and with what consequences. This includes paying close attention to evaluation across different governance levels (i.e. is evaluation done at EU level, in the Member States, or elsewhere, including the relationships between different evaluation actors?). The world is clearly anxious to evaluate whether or not it is on a development path that avoids catastrophic climate change (see EEA 2016). It is telling that the performance of wholly new governance initiatives such as the post-Paris Review Process (Christoff 2016) and EU energy governance ride on monitoring and evaluation exercises that are relatively novel and still finding their feet (Schoenefeld and Jordan 2017, Ringel and Knodt 2018). The currently uneven patterns of empirical insight result in part from very limited data availability; even very simple evaluation databases remain rare.

The fact that evaluation is on the rise in EU environmental governance, both theoretically and empirically, does not necessarily imply that this is unequivocally a normatively desirable development. Even though it is possible to document the rise of evaluation, research on its use, effects, and governance is only just emerging (see Schoenefeld and Jordan 2017). Furthermore, researchers could also engage with factors beyond the three arguably functional drivers considered here, including issues of power (Partzsch 2017) and the role of state (Duit et al. 2016) and non-state actors (Bäckstrand et al. 2017) in evaluation (Schoenefeld and Jordan 2017), as well as at the level of individual organisations (see Pattyn 2014). It would also be helpful to know how far the three core drivers and their relationships in the

EU environmental sector are applicable to other sectors and to other levels of governance. Comparative theoretical and empirical explorations that work across a range of different policy sectors could thus be very illuminating. In all these different ways, researchers stand to learn about environmental policy and politics by reflecting on policy evaluation.

Acknowledgments

We thank the special issue editors and two anonymous reviewers for very helpful comments on previous drafts. We presented earlier versions at ECPR Joint Sessions in Pisa in 2016, at the Regulatory Governance Conference in Tilburg in 2016, and at the University of Gothenburg in 2017. The members of the PPE group, including notably John Turnpenny (UEA) and the EKU Group (TU Darmstadt), provided helpful feedback.

Disclosure statement

No potential conflict of interest was reported by the authors.

Funding

JS and AJ benefitted financially from the COST Action INOGOV (IS1309). JS received financial support from a PhD studentship at UEA, and the German Federal Ministry of Education and Research (Reference: 03SFK4P0, Consortium ENavi, Kopernikus). AJ's contribution was supported by the 'UK in a Changing Europe Initiative' (ESRC PO 4030006272).

ORCID

Jonas J. Schoenefeld ⓘ http://orcid.org/0000-0002-9451-9174
Andrew J. Jordan ⓘ http://orcid.org/0000-0001-7678-1024

References

Adelle, C., Jordan, A., and Turnpenny, J., 2012. Proceeding in parallel or drifting apart? A systematic review of policy appraisal research and practices. *Environment and Planning C: Government and Policy*, 30 (3), 401–415.
Aldy, J.E., 2014. The crucial role of policy surveillance in international climate policy. *Climatic Change*, 126 (3–4), 279–292.
Aldy, J.E., 2018. Policy surveillance: its role in monitoring, reporting, evaluating and learning. *In*: A. Jordan, D. Huitema, H. van Asselt, and J. Forster, eds. *Governing climate change: polycentricity in action?* Cambridge: Cambridge University Press, 210–227.
Alkin, M.C. and Christie, C.A., 2004. An evaluation theory tree. *In*: M.C. Alkin, ed. *Evaluation roots: tracing theorists' views and influences*. Thousand Oaks: Sage, 12–65.

Bäckstrand, K., et al., 2017. Non-state actors in global climate governance: from Copenhagen to Paris and beyond. *Environmental Politics*, 26 (4), 561–579.

Batterbury, S.C., 2006. Principles and purposes of European Union cohesion policy evaluation. *Regional Studies*, 40 (2), 179–188.

Borrás, S. and Højlund, S., 2015. Evaluation and policy learning: the learners' perspective. *European Journal of Political Research*, 54 (1), 99–120.

Bovens, M., Hart, P., and Kuipers, S., 2006. The politics of policy evaluation. *In*: M. Moran, M. Rein, and R.E. Goodin, eds. *The Oxford handbook of public policy*. Oxford: Oxford University Press, 319–335.

Bovens, M., 2007. New forms of accountability and EU-governance. *Comparative European Politics*, 5 (1), 104–120.

Bovens, M., 2010. Two concepts of accountability: accountability as a virtue and as a mechanism. *West European Politics*, 33 (5), 946–967.

Bruyninckx, H., 2009. Environmental evaluation practices and the issue of scale. *New Directions for Evaluation*, 2009 (122), 31–39.

Chelimsky, E., 2006. The purposes of evaluation in a democratic society. *In*: I. Shaw, J. Greene, and M. Mark, eds. *The SAGE handbook of evaluation*. London: Sage Publications, 33–55.

Chelimsky, E., 2009. Integrating evaluation units into the political environment of government: the role of evaluation policy. *New Directions for Evaluation*, 123, 51–66.

Christoff, P., 2016. The promissory note: COP 21 and the Paris Climate Agreement. *Environmental Politics*, 25 (5), 765–787.

Collins, K. and Earnshaw, D., 1992. The implementation and enforcement of European community environment legislation. *Environmental Politics*, 1 (4), 213–249.

Crabbé, A. and Leroy, P., 2008. *The handbook of environmental policy evaluation*. London: Earthscan.

Dorsch, M.J. and Flachsland, C., 2017. A polycentric approach to global climate governance. *Global Environmental Politics*, 17 (2), 45–64.

Duit, A., Feindt, P.H., and Meadowcroft, J., 2016. Greening Leviathan: the rise of the environmental state? *Environmental Politics*, 25 (1), 1–23.

European Commission, 1996. *Evaluation: concrete steps towards best practice across the commission*. Brussels: European Commission.

European Environment Agency, 2016. *Environment and climate policy evaluation*. Copenhagen: European Environment Agency.

Feldman, D.L. and Wilt, C.A., 1996. Evaluating the implementation of state-level global climate change programs. *Journal of Environment and Development*, 5 (1), 46–72.

Hanberger, A., 2012. Framework for exploring the interplay of governance and evaluation. *Scandinavian Journal of Public Administration*, 16 (3), 9–27.

Haug, C., et al., 2010. Navigating the dilemmas of climate policy in Europe: evidence from policy evaluation studies. *Climatic Change*, 101 (3–4), 427–445.

Haug, C., 2015. *Unpacking learning: conceptualising and measuring the effects of two policy exercises on climate governance*. Thesis (PhD). Vrije Universiteit Amsterdam.

Hayward, R.J., et al., 2014. Evaluation under contract: government pressure and the production of policy research. *Public Administration*, 92 (1), 224–239.

Hermans, L.M., 2009. *A paradox of policy learning: evaluation, learning and accountability*. Available from: https://ecpr.eu/Filestore/PaperProposal/a3847a8d-4df2-4b74-a8a5-2fdb96d8fc48.pdf

Hertting, N. and Vedung, E., 2012. Purposes and criteria in network governance evaluation: how far does standard evaluation vocabulary takes us? *Evaluation*, 18 (1), 27–46.

Hildén, M., 2011. The evolution of climate policies – the role of learning and evaluations. *Journal of Cleaner Production*, 19 (16), 1798–1811.

Hildén, M., Jordan, A.J., and Rayner, T., 2014. Climate policy innovation: developing an evaluation perspective. *Environmental Politics*, 23 (5), 884–905.

Højlund, S., 2015. Evaluation in the European Commission. *European Journal of Risk Regulation*, 1, 35–46.

Howlett, M., 2014. Why are policy innovations rare and so often negative? Blame avoidance and problem denial in climate change policy-making. *Global Environmental Change*, 29, 395–403.

Huitema, D., et al., 2011. The evaluation of climate policy: theory and emerging practice in Europe. *Policy Sciences*, 44 (2), 179–198.

Jacob, S., Speer, S., and Furubo, J., 2015. The institutionalization of evaluation matters: updating the International Atlas of Evaluation 10 years later. *Evaluation*, 21 (1), 6–31.

Jordan, A.J., et al., 2015. Emergence of polycentric climate governance and its future prospects. *Nature Climate Change*, 5, 977–982.

Jordan, A.J., et al., 2018. Governing climate change polycentrically: setting the scene. *In*: A.J. Jordan, D. Huitema, H. van Asselt, and J. Forster, eds. *Governing climate change: polycentricity in action?* Cambridge: Cambridge University Press, 3–25.

Judge, D., 1992. A green dimension for the European community? *Environmental Politics*, 1 (4), 1–9.

Knaap, G.J. and Kim, T.J., 1998. *Environmental program evaluation: a primer*. Urbana: University of Illinois Press.

Knill, C. and Liefferink, D., 2013. The establishment of EU environmental policy. *In*: A. Jordan and C. Adelle, eds. *Environmental policy in the EU*. 3rd ed. London: Routledge, 13–31.

Majone, G., 1994. The rise of the regulatory state in Europe. *West European Politics*, 17 (3), 77–101.

Martens, M., 2010. Voice or loyalty? The evolution of the European Environment Agency (EEA). *Journal of Common Market Studies*, 48 (4), 881–901.

Mastenbroek, E., van Voorst, S., and Meuwese, A., 2016. Closing the regulatory cycle? A meta evaluation of ex-post legislative evaluations by the European Commission. *Journal of European Public Policy*, 23 (9), 1329–1348.

McAdam, D., 1996. Conceptual origins, current problems, future directions. *In*: D. McAdam, J.D. McCarthy, and M.N. Zald, eds. *Comparative perspectives on social movements: political opportunities, mobilizing structures, and cultural framings*. Cambridge: Cambridge University Press, 23–40.

Mickwitz, P., 2003. A framework for evaluating environmental policy instruments context and key concepts. *Evaluation*, 9 (4), 415–436.

Mickwitz, P., 2006. *Environmental policy evaluation: concepts and practice*. Thesis (PhD). University of Tampere.

Mickwitz, P., 2013. Policy evaluation. *In*: A. Jordan and C. Adelle, eds. *Environmental policy in the EU: actors, institutions and processes*. London: Routledge, 267–286.

OECD, 2001. *Evaluation feedback for effective learning and accountability*. Available from: https://www.oecd.org/dac/evaluation/2667326.pdf

Partzsch, L., 2017. 'Power with' and 'power to' in environmental politics and the transition to sustainability. *Environmental Politics*, 26 (2), 193–211.

Pattyn, V., 2014. Why organizations (do not) evaluate? Explaining evaluation activity through the lens of configurational comparative methods. *Evaluation*, 20 (3), 348-367.

Pattyn, V., et al., 2018. Policy evaluation in Europe. *In*: E. Ongaro and S. van Thiel, eds. *The Palgrave Handbook of Public Administration and Management in Europe*. London: Palgrave Macmillan, 577-593.

Pawson, R. and Tilley, N., 2014. *Realistic evaluation*. London: Sage.

Pleger, L. and Sager, F., 2016. Betterment, undermining, support and distortion: a heuristic model for the analysis of pressure on evaluators. *Evaluation and Program Planning*. doi:10.1016/j.evalprogplan.2016.09.002

Pollitt, C., 1998. Evaluation in Europe: boom or bubble? *Evaluation*, 4 (2), 214-224.

Radaelli, C.M., 2010. Rationality, power, management and symbols: four images of regulatory impact assessment. *Scandinavian Political Studies*, 33 (2), 164-188.

Radaelli, C.M. and Dunlop, C.A., 2013. Learning in the European Union: theoretical lenses and meta-theory. *Journal of European Public Policy*, 20 (6), 923-940.

Regeer, B.J., et al., 2016. Exploring ways to reconcile accountability and learning in the evaluation of niche experiments. *Evaluation*, 22 (1), 6-28.

Rhodes, R.A.W., 1996. The new governance: governing without government. *Political Studies*, 44 (4), 652-667.

Ringel, M. and Knodt, M., 2018. The governance of the European Energy Union: efficiency, effectiveness and acceptance of the Winter Package 2016. *Energy Policy*, 112, 209-220.

Sabel, C.F., 1994. Learning by monitoring: the institutions of economic development. *In*: N.J. Smelser and R. Swedberg, eds. *The handbook of economic sociology*. Princeton: Princeton University Press, 137-165.

Sanderson, I., 2002. Evaluation, policy learning and evidence-based policy making. *Public Administration*, 80 (1), 1-22.

Schattschneider, E.E., 1975. *The semi-sovereign people: a realist's view of democracy in America*. New York: Rinehart & Winston.

Schoenefeld, J.J., Hildén, M., and Jordan, A.J., 2018. The challenges of monitoring national climate policy: learning lessons from the EU. *Climate Policy*, 18 (1), 118-128.

Schoenefeld, J.J. and Jordan, A.J., 2017. Governing policy evaluation? Towards a new typology. *Evaluation*, 23 (3), 274-293.

Schrefler, L., 2016. *Evaluation in the European Commission: rolling check-list and state of play*. Brussels: European Parliamentary Research Service.

Stame, N., 2003. Evaluation and the policy context: the European experience. *Evaluation Journal of Australasia*, 3 (2), 37-43.

Stame, N., 2006. Governance, democracy and evaluation. *Evaluation*, 12 (1), 7-16.

Stame, N., 2008. The European project, federalism and evaluation. *Evaluation*, 14 (2), 117-140.

Stephenson, P., 2015. Reconciling audit and evaluation? The shift to performance and effectiveness at the European court of auditors. *European Journal of Risk Regulation*, 6 (1), 79-89.

Stern, E., 2009. Evaluation policy in the European Union and its institutions. *New Directions for Evaluation*, 2009 (123), 67-85.

Tosun, J. and Schoenefeld, J.J., 2017. Collective climate action and networked climate governance. *Wiley Interdisciplinary Reviews: Climate Change*, 8 (1). doi:10.1002/wcc.440

Toulemonde, J., 2000. Evaluation culture (s) in Europe: differences and convergence between national practices. *Vierteljahrshefte zur Wirtschaftsforschung*, 69 (3), 350–357.

Turnpenny, J., et al., 2016. Environment. *In*: C.A. Dunlop and C.M. Radaelli, eds. *Handbook of regulatory impact assessment*. Cheltenham: Edward Elgar, 193–208.

Uitto, J.I., 2016. Evaluating the environment as a global public good. *Evaluation*, 22 (1), 108–115.

UNFCCC, 2015. *Adoption of the Paris agreement*. Available from: http://unfccc.int/resource/docs/2015/cop21/eng/l09r01.pdf

van der Meer, F. and Edelenbos, J., 2006. Evaluation in multi-actor policy processes: accountability, learning and co-operation. *Evaluation*, 12 (2), 201–218.

van Voorst, S. and Mastenbroek, E., 2017. Enforcement tool or strategic instrument? The initiation of ex-post legislative evaluations by the European Commission. *European Union Politics*, 18 (4), 640–657.

Vedung, E., 1997. *Public policy and program evaluation*. New Bruswick, NJ: Transaction Publishers.

Vo, A.T. and Christie, C.A., 2015. Advancing research on evaluation through the study of context. *New Directions for Evaluation*, 2015 (148), 43–55.

Warren, P., 2014. A review of demand-side management policy in the UK. *Renewable and Sustainable Energy Reviews*, 29, 941–951.

Weiss, C.H., 1993. Where politics and evaluation research meet. *Evaluation Practice*, 14 (1), 93–106.

Weiss, C.H., 1999. The interface between evaluation and public policy. *Evaluation*, 5 (4), 468–486.

WWF, 2010. *Climate policy tracker for the European Union*. Brussels: World Wide Fund for Nature.

Zito, A.R., Burns, C., and Lenschow, A., 2019. Is the trajectory of European Union environmental policy less certain? *Environmental Politics*, 28 (2). [this issue].

Zito, A.R. and Schout, A., 2009. Learning theory reconsidered: EU integration theories and learning. *Journal of European Public Policy*, 16 (8), 1103–1123.

Zwaan, P., van Voorst, S., and Mastenbroek, E., 2016. Ex post legislative evaluation in the European Union: questioning the usage of evaluations as instruments for accountability. *International Review of Administrative Sciences*, 82 (4), 674–693.

Index

Abbreviations used in the index:
CoM - Covenant of Mayors
ENGO - environmental non-government organisations
EU - European Union
UK - United Kingdom
Page numbers in **bold** refer to material in tables; page numbers in *italics* refer to material in figures

Aarhus Convention (1998) 157, 165–166; national legal systems effects 167
accountability in policy evaluation 183–184
acid rain 5
Action for the Protection of the Environment in the Mediterranean Region (MEDSPA) 145
administrative courts in Sweden 170–171
air pollution legislation in Germany 170
Alkin, M.C. 183
Amsterdam Treaty (1997) 67
analytical focus 2–3
antagonism, policy evaluation drivers 189–190
Austria: accession to EU 131; ENGOs standing 166–167; green reputation 74; legislation non-compliance 141; material preclusion in law 169

Barroso Commission: climate objectives & energy policy 48–59; Renewable Energy Directive development 46
Basic Emission Inventories (BEIs) 110
BEIs (Basic Emission Inventories) 110
Berlin Infringement Database 133–134
Better Regulation Agenda 149–150
biofuels 50–51
Börzel, T. A. 87
Brexit 14, 85–106; de-Europeanisation 93–99; devolution settlements 92; disengagement and 87–88; disengagement as 99, **99**; EU-level policy dynamics 94–95; green policies 93; referendum campaign 93–97; remaining Member States, effects on 86
Brundtland Report (1993) 33
Bulgaria, legislation non-compliance 141
BUND 168
businesses, UK and EU environment agenda 91

C40 115
Cabinet and Directorate General for Climate Change (CLIMA) 42–43, 47; climate action mainstreaming 53, 54
carbon dioxide emissions 110; reduction 117, **118**
CBI (Confederation of British Industry) 95–96
CEES *see* Central and Eastern European States (CEES)
Central and Eastern European States (CEES): accession to EU 132; cognitive *vs.* exemplary leadership 80; financial support 145; support for 148; Western European Member states *vs.* 67, 75
centralised legislation enforcement 159–161
change: dimensions of 4–6; governments in 5–6; incremental 7; none 7; process dynamics of 6–9; theoretical approaches to 3–9, **4**
Chelimsky, E. 180, 183–184
Christie, C.A. 183
Christoff, P. 26
CJEU *see* European Court of Justice (CJEU)
Client Earth 97
Climate Alliance 115
climate change 26; biofuel impact 51; policies 109–113; priority lowering 57–58
Climate Change Act (UK) (2007) 88, 91, 116

climate policy integration (CPI) 42–61; co-benefit challenges 50–51; conditions for success 44–47, 55–58; definitions 45; effectiveness of 43; Environmental Policy Integration vs. 44–45
Climate Policy Tracker 185
Codd, J.A. 28
cognitive leadership 64–65, 75, **78**; European Council 69, 73
cognitive scenario, local authority engagement 112
Cohesion Fund 145
Collins, K. 156–157, 159
CoM *see* Covenant of Mayors (CoM)
Committee of Permanent Representatives (Coreper) 69–70, 71
Common Agricultural Policy 7
COMO (Covenant of Mayors Office) 114
complaints, legislation non-compliance 141
Confederation of British Industry (CBI) 95–96
consumers, stakeholders vs. 35–36
Copeland, P. 88
Coreper (Committee of Permanent Representatives) 69–70, 71
Council of Ministers: co-legislative powers with European Parliament 5; programmatic leadership 13–14
Council of the European Union 69–74; composition 69–70; leadership types 71–74
Council Working Groups 69–70, 71
Covenant of Mayors (CoM) 107–128; analytical framework of 113–114; carbon dioxide emissions 110; carbon dioxide reduction 117, **118**; definition 114–115; domestic contexts 121–123; implementation in UK 122; implementation of 116–120, **117**; information reliability 119; Italy vs. UK 112–113; local authorities common objectives 114–115; membership of 115–116; monitoring mechanisms **117**, 118–119, 120; municipal commitment **118**; SEAP implementation *117*; signatories 116, *117*; staff availability 120; training 119; trust & reputation 119; UK vs. Italy 122–123; validity of 123
Covenant of Mayors Office (COMO) 114
CPI *see* climate policy integration (CPI)
cultural dynamics 8–9
Czech Republic *see* Visegrad Group

de-centralised governance 25–26
de-centralised legislation enforcement 164–171; Aaarhus Convention 165–166; advantages 172; procedural autonomy 166–167; procedural right litigation 167–171
deep-core beliefs 45
de-Europeanisation 86, 87–88, **89**; Brexit referendum campaign 93–95; extent of 88; post-Brexit 95–96, 97–99; *see also* Brexit
de-modernisation process, Ecological Modernisation 24–25
Denmark: cognitive leadership 65; ENGOs standing 166–167; green reputation 74
desertions, CoM 122
destabilization 8–9
Deutsche Umwelthilfe 170
differentiated integration, member states 2
dimensions of change 4–6
Directorates General for Agriculture and Rural Development (DG AGRI) 47
Directorates General Transport and Energy (DG-TREN) 47
discourse analysis 28–29
discursive dynamics 4–5
disequilibria in integration 13
Djurgården-Lilla Värtans-Miljöskyddsförening 167–168
domestic concerns, Europeanisation 92
double depolitization 23; Ecological Modernisation 36–38

EAPs *see* Environmental Action Programmes (EAPs)
Earshaw, D. 156–157, 159
ECA (European Communities Act 1972) 93–94
Eckstein, H. 8–9
Ecological Modernisation (EM) 22–41; dominance of 36–7; within EU 26–28; rise of 24–26; weakness of 24
economic crisis 37, 46
economic growth: Ecological Modernisation and 25; environmental degradation and 34–35
economic resource, environmental protection as 32–33
Edelenbos, J. 190
EEA *see* European Environment Agency (EEA)
EEAS (European External Action Service) 68
Emissions Trading System (ETS) (2013) 13, 36, 42, 108
Energy Roadmap 2050 26–27
energy security, Ecological Modernisation 25

INDEX 201

enforcement: de-centralized legislation *see* de-centralised legislation enforcement; legislation non-compliance 138, 140–141, *142*, *143*, 144, 150
ENGOs *see* environmental non-governmental organisations (ENGOs)
entrepreneurial leadership 63–64, 75, **78**; European Council 68–69
ENVIREG (Regional Action Programme on the Initiative of the Commission Concerning the Environment) 145
Environmental Action Programmes (EAPs) 28–30, 192; analysis of 23; definition 28; discourse analysis 28–29, **31**; EAP1 (1973–1976) 30; EAP2 (1977–1981) 30, 32; EAP3 (1982–1986) 32–3; EAP4 (1987–1992) 33; EAP5 (1993–2000) 33–4; EAP6 (2000–2012) 34–5; EAP7 (2012–2020) 35–36; evaluation and 181; subject position frequency count **29**
environmental degradation, economic growth and 34–35
environmental legislation 131–138
environmental non-governmental organisations (ENGOs) 47, 86; Brexit, response to 95; court access 171; government structure 91; legislation enforcement 167–171; standing of 166–167
environmental policy 3, 9–15; establishment & early growth (1967–1984) 9–10; future developments (2000–) 12–15; integration and 1–2, 27; market & policy impulse reconciliation (1985–1999) 11; normal governance challenges (1999–2008) 11–12
Environmental Policy Integration (EPI) 67; climate policy integration *vs.* 44–45
Environment Council 70
Environment Working Group 70
EPI *see* Environmental Policy Integration (EPI)
ETS (Emissions Trading System) 13, 36, 42, 108
EUFJE (European Union Forum of Judges for the Environment) 147
European Commission: climate change 26–27; Covenant of Mayors Office (COMO) 114; Energy Security Strategy 115
European Communities Act (ECA) (1972) 93–94
European Council 65–69; cognitive leadership 69, 73; composition 66; entrepreneurial leadership 68–69; exemplary leadership 69, 73–74; leadership dynamics 80; leadership types 66–69; policy-making and 67–68; programmatic leadership 13–14; structural leadership 66–67, 71
European Court of Auditors 180
European Court of Justice (CJEU) 133; case numbers 162–163, *163*; legislation enforcement 158, *158*, 159; national procedures 166; time taken 160
European Environment Agency (EEA) 180; policy evaluation expectations 191–192
European External Action Service (EEAS) 68
Europeanisation 87–88, 90–92; policy 90; politics 90–91; polity 92
European Parliament 5
European Union Forum of Judges for the Environment (EUFJE) 147
European Union Justice Scoreboard 172
European Union Network for the Implementation and Enforcement of Environmental Law 5–6
European Union Pilot 163–164
European Union Withdrawal Act (EUWA) 93–94; laws without infrastructure 97
Eurozone crisis 53–55
exemplary leadership 64, 75, **79**; European Council 69, 73–74

Financial Instrument for the Environment (LIFE) 145
Finland: accession to EU 131; climate policy evaluation 186; green reputation 74; legislation non-compliance 141
formal letter of warning 133
France, private legislation enforcement 167

German Greens 5
Germany: air pollution legislation 170; ENGO legislation enforcement 168, 170; ENGOs standing 166–167; green reputation 74; material preclusion in law 169; private legislation enforcement 167
global environmental leader 15
global financial crisis, environmental policy 12
Gove, M. 93
governance, polycentric systems as 111
government regulations, Ecological Modernisation 25–26
Greece: accession to EU 131; ENGO court access 171; legislation non-compliance 140; private legislation enforcement 167; temporal patterns of legislation non-compliance 134; *see also* Southern problem

Green Development Network (GDN) 76
Green Diplomacy Network (GDN) 68
green economy 35
Greener UK campaign 95
green growth 37
Green Growth Group (GGG) 76, 80
greenhouse gas emissions (GHGE): CoM 114; reduction of 64–65
Green Party (UK) 91
green policies, Brexit 93
green-washing 51–52
growth of EU 130

Haas, E. 7
Hajer, M.A. 24
Haung, C. 185
Hey, C. 30
Hildebrand, P. 9
Hildén, M. 186
Højlund, S. 190–191
Hungary *see* Visegrad Group

ICLEI 115
ideational dynamics 4–5
Implementation and Enforcement of Environmental Law (IMPEL) 92, 147
incremental change 7
information reliability, CoM 119
infringement procedures, legislation compliance 132–133
institutionalisation, Ecological Modernisation 36
institutionalist perspectives 8
integration: challenges to 12–13; disequilibria in 13; environmental policy and 1–2, 27; uncertainties in 2
Intended Nationally Determined Contributions (INDCs) 67
intentional exemplary leadership 65
Intergovernmental Panel on Climate Change 56
Internal Market implementation 140
Internal Market Scoreboard 144
investigations, legislation non-compliance 140–141
Ireland: high costs of legal procedures 169; legislation enforcement 167
Italy: private legislation enforcement 167; UK *vs.* with CoM 112–113

Jacquot, S. 112
Joint Research Centre (JRC) 108
Judge, D. 28
Juncker Commission (2014) 52, 161
Juncker, J.-C. 161

Kahn, S. 95
Kyoto Protocol (1997) 11, 48

land use planning 90
Leaders' Agenda 68
leadership 62–84; programmatic leadership 13–14; types of 63–65, **77**–*79 see also* cognitive leadership; entrepreneurial leadership; exemplary leadership; structural leadership
learning, policy evaluation 184–186
learning output, CoM 121–122, 123–124
legal procedures: cost in UK 168–169; difficulties with 171; high cost of 171
legislation: amendment of 145, *146*; de-centralized enforcement *see* de-centralised legislation enforcement; decline in 80; development of 10; enforcement *see* legislation enforcement; environmental legislation 131–138; national implementation 156–157, 160; open infringement *162*, 162–163, *163*
legislation compliance 129–155; qualitative/quantitative measurement 132; *see also* legislation non-compliance
legislation enforcement 14–15, 156–178; centralised 159–161; Commission priorities 161–164, *162*; de-centralised *see* de-centralised legislation enforcement; before European Court of Justice 158, *158*; private means 160–161
legislation non-compliance 131, 138, 140–141, 144–145, 147–148; measurement of 132–134; reduction of 147–148; temporal patterns 134, *135*, *136*, *137*, 138, *139*; *see also* legislation compliance
legitimating scenario, local authority engagement 112
Lisbon Process (2000) 11, 27
Lisbon Treaty (2009) 80; European Council guidelines 66
local administration, CoM 121–122

Maastricht Treaty 144
Machin, A. 13
Make It Work Initiative 149–150
management, legislation non-compliance 138, 140, 144–145, *146*, 147, 150
March, J. 6
market-correcting policies 149
Mastenbroek, E. 179–180
Mayor Adept Initiative 115
MEDSPA (Action for the Protection of the Environment in the Mediterranean Region) 145

Member States 74–76; alliances 74–76; differentiated integration 2; leadership dynamics 80; structural leadership 81
MFF *see* Multiannual Financial Framework (MFF)
military power, structural leadership 64
monitoring mechanisms, CoM **117**, 118–119, 120
Multiannual Financial Framework (MFF) 46; climate action mainstreaming 53–54; expenditure for climate action 56–57
multi-level governance 14
municipal commitment, CoM **118**

national implementation of legislation 156–157, 160
Nature Protection Act 95
negotiation, entrepreneurial leadership 64
neofunctionalism 7–8
The Netherlands: cognitive leadership 65; green reputation 74
networking, policy evaluation 184
New Environmental Policy Instruments (NEPIs) 107
non-governmental organisations (NGOs): Europeanisation 90–91; legislation enforcement 157; *see also* environmental non-governmental organisations (ENGOs)
non-notification (non-communication), legislation 141
normative term, Ecological Modernisation 24
Nottingham Declaration Partnership on Climate Change 122

Olsen, J 6
Ostrom, E 111

Paris Agreement (UNFCCC 2015) 186
parliamentary questions, legislation non-compliance 141
party competition 5–6
petitions, legislation non-compliance 141
Poland *see* Visegrad Group
policy: Ecological Modernisation as 24–25; Europeanisation 90; implementation 14
policy actors 109–113
policy coalitions 45–46
policy-core beliefs 45; Renewable Energy Directive 48–50
policy entrepreneurs 45
policy evaluation 179–198; across government levels 193–194; complexity of 192–193; drivers of 183–192, *189*, 193; emergence of 181–183, **182**; institution outputs evaluation 187–188; neglect of 180
policy framing 5
policy-making 67–68
policy paradigm 24
policy text deconstruction 28
political opportunity, policy evaluation 186–188
politics, Europeanisation 90–91
polity, Europeanisation 92
polycentric governance 184
polycentric systems: CoM 124; governance as 111
Portugal: accession to EU 131; legislation non-compliance 140; private legislation enforcement 167; temporal patterns of legislation non-compliance 134; *see also* Southern problem
procedural autonomy, legislation enforcement 166–167
procedural right litigation 167–171
process dynamics of change 6–9
programmatic leadership 13–14
punctuated equilibrium concept 8

reasoned opinions 133
Regional Action Programme on the Initiative of the Commission Concerning the Environment (ENVIREG) 145
Registration, Evaluation and Authorisation of Chemicals (REACH) 98
regulation, pressure for flexibility 12
Regulatory Fitness and Performance Programme (REFIT) 72–73, 161
Renewable Electricity Directive (2001) 48–49
renewable energy 48–49; climate mitigation 56
Renewable Energy Directive (RED) 13, 15, 43–44, 55–56; adoption 48, 49–50; development 46–47, 48–50; electricity/heating 48; negotiation of 48–50
Renewable Energy Road map (2007) 49
Research Service of the European Parliament 191
Rio Declaration (1992) 164
Risse, T. 87
Romania, legislation non-compliance 141

Sabel, C.F. 190
scope of conflict 5
SD *see* Sustainable Development (SD)
SEAPs *see* Sustainable Energy Action Plans (SEAPs)
Single European Act (SEA, 1986) 11, 66–67, 72, *72*

Single Market, national legislation harmonization 10
Slovakia: bear case 169–171; ENGOs 169–170; *see also* Visegrad Group
Southern problem 131, 132, 148–149; legislation non-compliance 134, 138
Spain: accession to EU 131; ENGO court access 171; legislation non-compliance 140; private legislation enforcement 167; temporal patterns of legislation non-compliance 134; *see also* Southern problem
SSNC (Swedish Society for Nature Conservation) 170–171
stakeholder, consumer *vs.* 35–36
strategic scenario, local authority engagement 112
Strategy Europe 2020 107–108
structural leadership 63, 64, **77**; European Council 66–67, 71; Member States 81
structuration, Ecological Modernisation 36
Sustainable Development (SD) 11; Environmental Action Programmes 33
Sustainable Energy Action Plans (SEAPs) 108; implementation of 115, *117*; indicators for 110
Sweden: accession to EU 131; administrative courts 170–171; cognitive leadership 65; ENGO legislation enforcement 167–168; ENGOs standing 166–167; green reputation 74; legislation non-compliance 141; Slovak bears case 170–171
Swedish Environmental Protection Agency 170–171
Swedish Society for Nature Conservation (SSNC) 170–171
symbolic scenarios, local authority engagement 112

Technical Assistance Information Exchange Office (TAIEX) 145, 147
Thompson, M. 6
Torrey Canyon disaster (1967) 9–10
trade globalisation 27
transnational municipal networks (TMNs) 108; studies on climate 110–111
transnational systems 6
Trump, D. 5
25-year environment plan (28YEP) (2018) 95

unintentional exemplary leadership 65
United Kingdom (UK): ENGO legislation enforcement 168–169; Europeanisation *see* Europeanisation; Green Growth Group (GGG) 76; high costs of legal procedures 168–169; Italy *vs.* with CoM 112–113; leaving EU *see* Brexit; legislation enforcement 167
United Nations Framework Convention on Climate Change (UNFCCC) 43
UN Paris Climate Conference (2015) 67
UN Stockholm Conference (1972) 66

van der Meer, F. 190
Vedung, E. 180
Visegrad Group 75–76; environmental priorities 80
VLK Lesoochranárske zoskupenie 169–170

Weale, A. 22–23
Weiss, C.H. 185
Western European Member states, Central and Eastern European States *vs.* 67, 75
Woll, C. 112
World Wildlife Fund (WWF) 185
Wurzel, R. 13–14